Springer Theses

Recognizing Outstanding Ph.D. Research

Aims and Scope

The series "Springer Theses" brings together a selection of the very best Ph.D. theses from around the world and across the physical sciences. Nominated and endorsed by two recognized specialists, each published volume has been selected for its scientific excellence and the high impact of its contents for the pertinent field of research. For greater accessibility to non-specialists, the published versions include an extended introduction, as well as a foreword by the student's supervisor explaining the special relevance of the work for the field. As a whole, the series will provide a valuable resource both for newcomers to the research fields described, and for other scientists seeking detailed background information on special questions. Finally, it provides an accredited documentation of the valuable contributions made by today's younger generation of scientists.

Theses are accepted into the series by invited nomination only and must fulfill all of the following criteria

- They must be written in good English.
- The topic should fall within the confines of Chemistry, Physics, Earth Sciences, Engineering and related interdisciplinary fields such as Materials, Nanoscience, Chemical Engineering, Complex Systems and Biophysics.
- The work reported in the thesis must represent a significant scientific advance.
- If the thesis includes previously published material, permission to reproduce this must be gained from the respective copyright holder.
- They must have been examined and passed during the 12 months prior to nomination.
- Each thesis should include a foreword by the supervisor outlining the significance of its content.
- The theses should have a clearly defined structure including an introduction accessible to scientists not expert in that particular field.

More information about this series at http://www.springer.com/series/8790

Adam Iaizzi

Magnetic Field Effects in Low-Dimensional Quantum Magnets

Doctoral Thesis accepted by Boston University, MA, USA

Springer

Adam Iaizzi
Department of Physics
National Taiwan University
Taipei, Taiwan

ISSN 2190-5053 ISSN 2190-5061 (electronic)
Springer Theses
ISBN 978-3-030-01802-3 ISBN 978-3-030-01803-0 (eBook)
https://doi.org/10.1007/978-3-030-01803-0

Library of Congress Control Number: 2018960268

This Springer imprint is published by the registered company Springer Nature Switzerland AG
The registered company address is: Gewerbestrasse 11, 6330 Cham, Switzerland

... science is more than a body of knowledge. It's a way of thinking, a way of skeptically interrogating the universe with a fine understanding of human fallibility. If we are not able to ask skeptical questions, to interrogate those who tell us that something is true, to be skeptical of those in authority, then we're up for grabs for the next charlatan, political or religious, who comes ambling along.

Carl Sagan

Timestamp 3:52 from this May 1996 interview with Charlie Rose
https://charlierose.com/videos/9094.

This thesis is dedicated to:
Dr. Arianna Wright Rosenbluth
The author of the first-ever implementation
of Metropolis Algorithm Monte Carlo

Supervisor's Foreword

A goal of computational materials science is to reproduce or predict physical properties as precisely as possible. However, in many cases it is impossible to include all interactions and quantum mechanical effects and still be able to carry out reliable calculations. From the standpoint of fundamental condensed matter physics, many details pertaining to specific materials are mere distractions, and one would instead like to construct the simplest possible models to describe physical phenomena of interest. A successful approach to studying quantum phases of matter and the phase transitions taking place between them is to construct field theories that are only constrained by relevant symmetries. If a given field theory can be solved, or treated with sufficiently good approximations, a wealth of information can be extracted on possible ground states and their excitations. However, solving quantum field theories is also no easy task, and it is not always possible to draw definite conclusions. Another approach is to formulate simplified lattice models inspired by real materials but not using the exact lattice structure and interactions of any specific material. If successful, these models (Hamiltonians) host the low-energy quantum states of interest, and they can sometimes be characterized in an approximation-free manner by numerical simulations. Since the low-energy physical properties of the lattice models should correspond exactly to some quantum field theory, direct connections can be made with theory as well as with experiments probing generic features.

Even simplified lattice models are not automatically amenable to efficient numerical simulations. In particular, Monte Carlo simulations of quantum systems are often hampered by mixed signs of the weight function characterizing the configuration space—the infamous sign problem. Models tailored both from the perspective of interesting physics and computational tractability have been termed "designer Hamiltonians"[1] and are playing an increasingly important role in quantum many-body physics. Adam Iaizzi's thesis is devoted to constructing and studying a class of designer Hamiltonians in the field of quantum magnetism.

[1]R.K. Kaul, R.G. Melko, and A.W. Sandvik, Annu. Rev. Condens. Matter Phys. **4**, 179 (2013).

The work starts from the J-Q family of $S = 1/2$ quantum spin models on the 2D square lattice.[2] These models contain the common Heisenberg exchange interaction of strength J and a multi-spin interaction of strength Q. When $Q = 0$ the system has an antiferromagnetic ground state. The Q interaction favors the formation of spatially correlated singlets, and, above a critical value of Q/J, a spontaneously dimerized ground state. The transition separating the antiferromagnetic and dimerized phases is an intriguing "deconfined" quantum critical point, at which the excitations become fractionalized and exhibit unusual behaviors normally not found in 2D antiferromagnets. The J-Q model has been extensively studied in this context. Iaizzi expands on the J-Q model in both one and two dimensions by adding an external magnetic field, thereby accessing a broader range of physical phenomena. The quantum Monte Carlo algorithms, which can be adapted to many other models as well, are described in detail in this thesis and many new results are presented.

Even in the conventional ($Q = 0$) Heisenberg model the field effects have not previously been fully characterized; the Q interactions introduce further rich behaviors that mimic some of the properties expected in traditional frustrated quantum magnets (which are difficult to study because of the aforementioned sign problems). Iaizzi studies the experimentally important transition into a fully polarized ferromagnetic state and discovers a case of *metamagnetism*—a discontinuous jump in the magnetization versus field above a threshold value of Q/J. This threshold value itself corresponds to a kind of quantum criticality, which is here characterized quantitatively, including an exact analytical solution in one dimension. A previous theory for the continuous transition into full polarization for lower Q/J is tested in detail in both one and two dimensions. At weaker fields and Q/J close to the deconfined critical point, it is interesting to study the consequences of the spinons (the putative deconfined excitations) induced by the field in two dimensions. Iaizzi's work here aims to resolve the thermodynamic behaviors distinguishing spinons from the conventional magnon (spin waves) excitations. In addition to solving some key problems, the thesis points to intriguing open questions that could be addressed in future work.

Professor of Physics
Boston University, Boston, MA, USA

Anders W. Sandvik

[2] A.W. Sandvik, Phys. Rev. Lett. **98**, 227202 (2007).

Acknowledgments

My work was supported by the NSF under grants No. DMR-1710170 and DMR-1410126. The collaboration with Kedar Damle at the Tata Institute of Fundamental Research in Mumbai was supported by a travel grant from the APS-IUSSTF Physics PhD Student Visitation Program. My travel for presentations and other collaboration was supported by the NSF, the Simons Foundation, the Boston University Graduate Student Organization, the American Physical Society, and the Computational Science Research Center—Beijing (Natural Science Foundation of China Grant No. U1530401). The computational work reported in this paper was performed in part on the Shared Computing Cluster administered by Boston University's Research Computing Services.

Obviously I must thank my advisor Anders Sandvik for his patience and guidance and for providing me with so many opportunities to grow as a scientist over the past 5 years. I also must thank David Campbell for introducing me to quantum many-body physics, for his mentorship as I navigated graduate school and postdoc applications, and for nominating my thesis for a Springer Thesis Award. I am deeply grateful to my collaborators Kedar Damle and Dries Sels for their assistance and insights. I owe a great thanks to my collaborators Harley D. Scammell and Oleg P. Sushkov for their help with Chap. 4. In addition, I would like to thank the following people for useful conversations: Cenke Xu, Hai-Qing Lin, Sylvain Capponi, Ling Wang, Oleg Sushkov, and Harley Scammell. A special thanks to Hong Gang Luo, Bin-Bin Mao, and Chen Cheng for alerting me to an error in one of my preprints. I would also like to thank all members of the Sandvik group at Boston University, especially Phillip Weinberg, Pranay Patil, Na Xu, Thomas Lang, Maro Suwa, Nvsen Ma, Tim Khouw, and Ying Tang. Thanks also to my readers Anders Sandvik and David Campbell, along with the rest of my committee: Claudio Chamon, Rob Carey, and Shyam Erramilli.

I owe a great debt to my family, especially my parents, Tracey Iaizzi and Garry Iaizzi, and my grandmother, Marion Heider. And also to my friends, whose kindness, compassion, and homework help have kept me sane throughout graduate school. Last and certainly not least, I would like to thank my wife, Vanessa Calaban, for her love and support throughout this seven-year process, all the way from applying to graduate school to finding a postdoc and writing this very document; I cannot imagine having a better partner in my life.

Taipei, Taiwan

Adam Iaizzi
iaizzi@bu.edu
www.iaizzi.me
https://orcid.org/0000-0001-5570-8282

Contents

Parts of this thesis have been published in the following journal articles:

1. Adam Iaizzi, Harley D. Scammell, Oleg P. Sushkov, and Anders W. Sandvik. Direct numerical observation of Bose-Einstein condensation of spinons (in preparation)
2. Adam Iaizzi, Kedar Damle, and Anders W. Sandvik. Metamagnetism and zero-scale-factor universality in the two-dimensional J-Q model. Physical Review B **98**, 064405 (2018). https://doi.org/10.1103/PhysRevB.98.064405.
3. Adam Iaizzi, Kedar Damle, and Anders W. Sandvik. Field-driven quantum phase transitions in $S = \frac{1}{2}$ spin chains. Physical Review B **95**, 174436 (2017). https://doi.org/10.1103/PhysRevB.95.174436
4. Adam Iaizzi and Anders W. Sandvik. 1D valence bond solids in a magnetic field. Journal of Physics: Conference Series **640**(1), 012043 (2015). https://doi.org/10.1088/1742-6596/640/1/012043

List of Abbreviations

AFM	Antiferromagnet(ic)
BEC	Bose–Einstein condensate/condensation
BKT (or KT)	Berezinskii–Kosterlitz–Thouless (transition)
DQC	Deconfined quantum critical(ity)
FM	Ferromagnet(ic)
LRO	Long-range order
QLRO	Quasi-long-range order
QMC	Quantum Monte Carlo
SSE	Stochastic series expansion
VBS	Valence-bond solid

Chapter 1
Introduction

This dissertation describes a study of phase transitions in low-dimensional quantum magnets in the presence of external fields using numerical methods, chiefly stochastic series expansion quantum Monte Carlo with directed loop updates and quantum replica exchange. I have taken special care to describe the historical and scientific context for both the topics and the numerical methods used. I focus on the J-Q model, a quantum many-body Hamiltonian acting on a lattice of localized spin-half degrees of freedom which augments the Heisenberg exchange, $J\vec{S}_i \cdot \vec{S}_j$, with a four-spin interaction of strength Q [1]. The Q term serves as a competing interaction which mimics many of the phenomena present in frustrated systems, but without the infamous "sign problem" that makes frustrated systems inaccessible to quantum Monte Carlo, the most reliable method for large-scale simulations [1, 2]. The J-Q model has been extensively studied at zero field, where the Q term drives a quantum phase transition from a Néel antiferromagnet (AFM) state to a valence-bond solid (VBS, a non-magnetic state consisting of a long-range-ordered arrangement of local singlet bonds between sites) [1, 3–10]. I extend previous work on the J-Q model by adding an external magnetic field, producing the J-Q-h model. The three body chapters of this describe detailed studies of the J-Q model in both one and two dimensions.

In 1D, I show that the Q term produces metamagnetism: jumps in the magnetization which previously were known to occur only in systems with frustration or intrinsic anisotropy [11, 12]. I derive an exact solution for the minimum coupling ratio $(Q/J)_{\min} = 2/9$ for metamagnetism to occur and show that the transition is caused by the onset of attractive interactions between magnons (spin flips against a polarized background) [12]. Below $(Q/J)_{\min}$, the saturation transition is continuous and it is governed by zero-scale-factor universality [12, 13].

In two dimensions, I also find metamagnetism above a critical coupling ratio $(Q/J)_{\min} \approx 0.417$, caused by the same mechanism as in the one-dimensional case [14]. Two dimensions is the upper critical dimension of the zero-scale-factor universality, so the continuous saturation transition is still governed by zero-scale-

A. Iaizzi, *Magnetic Field Effects in Low-Dimensional Quantum Magnets*,
Springer Theses, https://doi.org/10.1007/978-3-030-01803-0_1

factor universality, but with multiplicative logarithmic violations of the scaling. I find that these violations do not match the expected form [13].

In two dimensions I also use the field to explore new aspects of the deconfined quantum critical point, showing evidence for the existence of deconfined spinons, and describing a field-induced BKT transition. Finally, we turn our attention to the region around the Néel-VBS transition in the 2D J-Q model. This transition is interesting because it violates the Ginzburg–Landau paradigm which requires that direct transitions between phases breaking unrelated symmetries[1] be first order [15, 16]. Ample numerical evidence has now established that the Néel-VBS transition in the 2D J-Q model is *continuous* [1, 4, 5, 9, 10, 17–19]. The solution to this apparent contradiction is *deconfined quantum criticality*, where the critical point is described by exotic fractionalized excitations that are confined (like quarks in a proton) in both ordered phases [15, 16]. The deconfined fractionalized excitations at the Néel-VBS transition are spinons—spin-half bosons. Using the field, I show evidence of an anomalous temperature dependence of specific heat arising from field-induced gas of spinons at the deconfined quantum critical point and discuss the effects of a field-induced Berezinskii–Kosterlitz–Thouless-like transition.

1.1 How to Read This Dissertation

In the rest of this introductory chapter I will describe the field of condensed matter physics (focusing on quantum many-body physics, quantum magnetism, and quantum phase transitions) along with the nature and history of the computational tools used in this dissertation. This introduction provides motivation and background information designed to be helpful for the reader uninitiated in this field. Chapters 2 to 4 comprise the body of this dissertation. Each of them is written as a self-contained paper, with its own introduction, background information, and conclusions. These chapters can be read in any order, and any points where they rely on knowledge from a previous chapter it is referenced appropriately. In Chap. 2 we discuss metamagnetism and the saturation transition in the J-Q chain.[2] Chapter 3 covers the same subject matter, but for the 2D case.[3] In Chap. 4, I discuss the behavior of the J-Q model with a magnetic field near the deconfined quantum critical point, showing direct evidence of a gas of deconfined spinons.[4] In Chap. 5 I describe the theory and practice of the quantum Monte Carlo methods used to

[1]The Néel state breaks $O(3)$ rotational symmetry and the VBS breaks Z_4 lattice symmetry in 2D.

[2]Chapter 2 is a lightly edited version of my paper "Field-driven quantum phase transitions in $\mathbf{S} = \frac{1}{2}$ spin chains" appearing in [12].

[3]A slimmed-down version of Chap. 3 has been published in [14].

[4]The data presented in Chap. 4 is now being reanalyzed in collaboration with Harley D. Scammell and Oleg P. Sushkov, and a manuscript is in preparation [20].

conduct this work. It is possible to understand Chaps. 2 to 4 without knowing the details of these methods; the intention of Chap. 5 is instead to be pedagogically useful for the reader who is attempting to develop their own QMC program. Finally, in Chap. 6 I summarize the previous chapters and offer some brief concluding remarks.

1.2 What Is Computational Physics?

Physics is somewhat unusual among the sciences for its long-standing division into separate disciplines of theory and experiment. This division arose out of necessity as experiments and theory each grew so complicated that it became impossible for any individual to master both. Computational physics is the use of numerical methods to study physical systems. Today almost all physicists use at least some numerical methods, but many methods are sufficiently complex and subtle that they require specialists dedicated to their development and use—computational physicists. Computational physics is thus neither theory nor experiment, but serves as an integral part of both, a bridge between them, and a third branch in its own right.[5] Numerical methods step in where analytical methods fail, or where experiments are impossible. They can be simple integrators implemented in a few lines of code, massive 100,000-line commercial quantum chemistry packages, or anything in between. A few examples of numerical methods include numerical integration, matrix diagonalization, density functional theory, machine learning, and Monte Carlo.

Hereafter I will focus almost exclusively on quantum Monte Carlo (the workhorse of this document), a way of studying quantum systems through a mapping onto a classical problem (these methods are described in detail in Chap. 5). Quantum Monte Carlo is just one of a wide array of techniques that fall into the category of Monte Carlo which share at their core a reliance on *stochastic sampling*, i.e., random numbers. Monte Carlo is powerful because it provides *unbiased* results for large systems without uncontrolled approximations. Here the term unbiased means that the answer is without systematic error.[6] More direct analytical or numerical approaches often utilize uncontrolled approximations (like first-order perturbation theory) or are limited to extremely small systems (like exact diagonalization). In the former case, the price of these exact solutions is that they are an exact solution to an *approximation*, so the answer always includes

[5]Perhaps the clearest way to distinguish between a "computational theorist" and a "computational experimentalist" is whether their simulations have units.

[6]Formally, an unbiased estimator in which the expectation value is equal to the true value of the parameter to be measured [21, p. 135]. An example of a *biased* estimator would be a variational solution for the energy, in which case the answer is guaranteed to be greater than or equal to the true ground state energy.

some *systematic* error and in the latter case, the solutions are numerically exact,[7] but are limited to systems that are too small to answer many questions. Monte Carlo enables large-scale unbiased simulations of interacting many-body systems without relying on clever approximations. Monte Carlo does introduce random error, but this error is quantifiable (the error bars are known) and controlled (it can be reduced by simply collecting more data). Monte Carlo provides a sort of numerical experiment, transcending the limitations of chemistry and materials science to enable studies of any Hamiltonian that theorists can cook up.[8] The J-Q model used in this work has no obvious physical realization [1]; it is instead a "designer Hamiltonian" designed to study specific physics: deconfined quantum criticality. Even the cherished Heisenberg model, although it has close experimental analogues, is of course a fiction, a platonic ideal existing only in the minds of those who study it. Numerical experiments using quantum Monte Carlo are thus the only experiments that can be used to study the physics of these idealized models.

1.2.1 A Brief History of Computational Physics

Computational physics is older than one would expect given that computers themselves (in the modern sense of the word) are barely 70 years old. The beginnings of computational physics can be traced back to the very beginnings of physics itself. A full description of the history of computing is far beyond the scope of this dissertation. I will instead attempt to highlight some of the key developments that laid the foundation for modern Monte Carlo methods that I have relied upon to conduct the research presented here. In the early days, a computer was not a machine, but a person, a person who performed calculations laboriously and by hand (or with the help of mechanical calculators) [22]. The first example of an organized substantial calculation performed using multiple people might be the ill-fated attempt to verify Newton's theory of gravity by precisely predicting the date of the 1758 return of Halley's Comet [22, p. 16]. There is no closed-form solution for the motion of the comet that can account for the gravitational influence of all the planets. Instead, Alexis-Claude Clairaut set out to predict its position *numerically* (a controversial proposal), dividing the work between himself and two friends: Mr. Joseph Lalande[9] and Ms. Nicole-Reine Lepaute[10] [22, p. 16]. With the benefit of hindsight we know they had little hope of producing an accurate prediction, not due to any flaws in their approach, but due to missing information. The gas giants Uranus

[7]Numerically exact means that the answer is exact to the limits of machine precision and suffers from no other random or systematic error.

[8]Provided of course that it is Marshall positive, see Sect. 5.3.

[9]Full name: Joseph-Jérôme Le Français de Lalande.

[10]Full name: Nicole-Reine Étable de la Brière Lepaute.

and Neptune, which substantially influence the result, were unknown to science when Clairaut, Lalande, and Lepaute began their work [22, p. 23].

By the outbreak of World War II, artillery had developed sufficient range and accuracy that calculating artillery trajectories had become a complex task. Artillery targeting relied on precomputed "firing tables" with thousands of entries accounting for distance and other factors [23]. This increased need for computing came at a time when men were in short supply, so human computers were very often women.[11] Each entry in these firing tables would take a human computer days to complete, so human computers worked alongside analog mechanical computers such as the Differential Analyzer to accelerate their work [23]. The first digital electronic computer, the ENIAC (Electronic Numerical Integrator and Computer), was built at the University of Pennsylvania to further accelerate these calculations [23, 24]. The first programmers of the first electronic computer were the human computers that its descendants would eventually replace—the women of the Philadelphia Computing Section. The ENIAC was completed too late to assist in the war effort. Its first real world test was running a calculation on the hydrogen bomb in late 1945 [24].

These new electronic computers were not merely faster than their predecessors, but more flexible as well. Instead of being purpose-built for performing specific calculations they could be programmed—adapted to perform any computation desired. As time went on, people began to use digital computers for entirely new methods that have no pencil-and-paper analog. These "numerical experiments" opened up whole new areas of physics that were previously off-limits. One of the first such methods of "numerical experiment" was Monte Carlo, invented the same year the ENIAC was unveiled to the public [24, 25]. Monte Carlo was named not for a physicist, but after the famous casino.[12] The name is appropriate: the distinguishing characteristic of Monte Carlo is its reliance on stochastic sampling (i.e., use of random numbers).[13]

1.2.2 Development of the Metropolis Algorithm

The Metropolis Algorithm was developed at Los Alamos by scientists working on the liquid–solid transition of interacting hard disks using the newly constructed programmable electronic computer called MANIAC (Mathematical Analyzer Numerical Integrator and Calculator) [26]. Since they were interested only in equilibrium properties, there was no need to follow physical dynamics of the system and

[11] Women had previously served as human computers in other contexts as well, for example, the group of all-female computers at the Harvard Observatory [23] and of course the aforementioned Nicole-Reine Lepaute who worked on the Halley's Comet prediction [22, p. 16].

[12] The story goes that one of the inventors of Monte Carlo, Stan Ulam, named it for his uncle's proclivity for gambling [25, 27].

[13] The risk-averse reader can rest assured: the physicists in this casino metaphor are the house, not the gambler.

extract results from time averages. Instead, they could draw configurations (of the disks) from the equilibrium distribution and rely on ensemble averages. They accomplished this by stochastically modifying a random initial configuration over the course of many steps utilizing the following procedure for each step: (1) propose a "pseudo-move" (some suggested change to the system) and (2) accept this change if it lowers the energy *or* accept it with probability

$$P = e^{-\Delta E/T} \tag{1.1}$$

if it *increases* the energy. This produces a (Markov) chain of configurations drawn from the equilibrium distribution. Here I will take some time to discuss the genesis of the Metropolis–Hastings Algorithm and leave the detailed description of the theory behind this method to Chap. 5.

Early Monte Carlo simulations were focused on solving particular problems relevant to weapons design where the transition probabilities were fixed by known physical properties (decays, cross sections, etc.) and the output of the simulation was the asymptotic distribution produced by these transitions [25]. The Metropolis Algorithm was a major breakthrough that inverted this problem, allowing simulations based on a specified asymptotic distribution (like the Boltzmann distribution), with the transition probabilities being a creation of the simulation designer (subject to detailed balance) [25]. Thus, the Metropolis Algorithm was a general method for determining the equilibrium thermodynamic properties of any classical system [25] and its invention was not merely an extension of a previous technique but a distinctly creative act resulting in a wholly new method [25, p. 14].

The Metropolis Algorithm first appeared in a 1953 paper [26] authored by Nicholas Metropolis, Marshall Rosenbluth, Arianna Wright Rosenbluth, Augusta Teller, and Edward Teller. The algorithm is named for the first author, Nicholas Metropolis. Despite his appearance as first author, Metropolis is said to have made no scientific contribution to the paper [28].[14] Most of the work was done by Marshall and Arianna Rosenbluth, with the key insight of using ensemble averages coming from Edward Teller. The first full computer implementation of this revolutionary algorithm was programmed entirely by Arianna Wright Rosenbluth (although Augusta Teller had done some preliminary work) [25, 27–29]. It took some time for the Metropolis Algorithm to become widely used due to a combination of factors including limited availability of computers and a reluctance to accept the use of numerical methods in theoretical physics [25]. Its adoption received a boost in 1970 when it was further generalized by Hastings [30] and as a result it is also known as the Metropolis–Hastings Algorithm. The Metropolis Algorithm has since become the most common form of Monte Carlo and spread beyond physics to

[14] Although here we rely on an interview with Marshall Rosenbluth 50 years after the events, this conclusion is supported by the absence of almost any mention of the algorithm in Metropolis' later publications or his memoirs [25], an absence that is remarkable given that the algorithm bears his name. Metropolis did mention the algorithm in a 1987 article in *Los Alamos Science* [31, p. 129], he did not explicitly claim to have been part of the team that invented it.

chemistry, biology, social science, finance, and even pure math [27]; its use is now so widespread that it is commonly mistaken for being a synonym for Monte Carlo itself [25].

Despite its importance, the origin of the algorithm was nearly lost to history. The adoption of the Metropolis algorithm was initially very slow [25]. By the time the algorithm became widely used, the original authors had all moved on to other things.[15] The credit for preserving the history of this algorithm can be given in large part to J.E. Gubernatis, who invited Marshall Rosenbluth to a conference celebrating the 50th anniversary of the Metropolis paper in the final year of his (Rosenbluth's) life [25, 29], interviewed Marshall and the other living author, Arianna Wright Rosenbluth [25, 27], and recorded some second-hand recollections of conversations with Edward Teller [25].

1.2.3 Toward a More Detailed Balance

Few lay people could name a famous woman physicist besides Marie Curie. In an extremely informal survey of my department, I found that few of my colleagues could name more than one other famous woman physicist.[16] Of course, physics has a very real problem in this regard: women are grossly underrepresented in physics. Only 10% of physics faculty at physics degree-granting institutions are women and more than 20% of these departments have *zero* women faculty [32].[17] In 2015, only about 20% of bachelor's degrees and doctoral degrees in physics were awarded to women (see Fig. 1.1) [33, 34]. In recruiting and retaining women, physics falls behind every other scientific field except for engineering. Worse yet, in the top panel of Fig. 1.1, we can see that the percentage of physics bachelor's degrees awarded to women has actually *fallen* over the past decade. Most of the missing women leave physics between high school (where about 50% of physics students are women [32]) and earning a bachelor's degree, after which point women stay in physics at roughly the same rate as men [32]. There is a litany of systemic issues that drive women away from physics and make it difficult for those who remain to succeed.[18] A full discussion of these is well beyond the scope of this dissertation (or indeed, a single dissertation). One of the most commonly supposed reasons for this attrition is the lack of representation of women in physics: young women do not see themselves as

[15] The authors could be forgiven for this oversight, given that their other accomplishments (such as their roles in developing the hydrogen bomb) had more immediately obvious applications.

[16] I specifically asked my colleagues to name someone famous that the general public might know, not women physicists they know personally or well-known women in their subfield.

[17] These numbers are especially old, dating from a 2002 survey, but the turnover of faculty positions is low, so any change since then is likely to be small.

[18] For example, women are more likely to suffer from imposter syndrome and more likely to have a poor relationship with their advisor [35] and their papers are cited fewer times than comparable papers authored by men [36, 37].

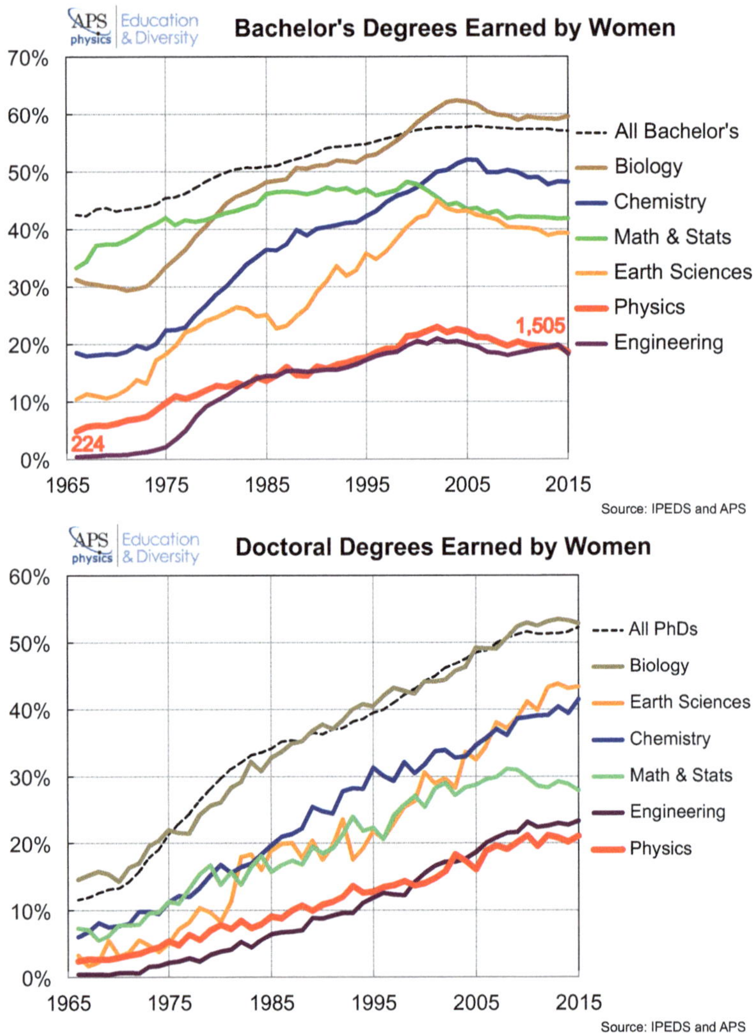

Fig. 1.1 (Top) women as a percentage of total bachelor's degree recipients by field. Credit: APS/Source: IPEDS Completion Survey [33]. (Bottom) women as a percentage of total PhD recipients by field. Credit: APS/Source: IPEDS Completion Survey [34] (both used with permission)

physicists because the public faces of physics are overwhelmingly white and male. Obviously, part of the representation problem is that there are not enough women physicists. As a discipline, we need not compound this problem by systematically failing to celebrate the very real contributions that women have made (and continue to make) to physics. Women's contributions to physics have been systematically ignored, or their credit allocated to men. Here I will briefly highlight some of the accomplishments of women in computational physics.

Despite the obstacles, there have always been women computational physicists (from literally the very beginning with Nicole-Reine Lepaute [22, p. 16]). Human computers were very often women; those positions offered a rare opportunity for women to work in mathematical and scientific fields at a time when few such opportunities existed. This was especially true during the second World War, where women played a critical role in the war effort as computers and mathematicians. Human computing was not menial labor, it required immense talent and training. In spite of this, human computers were typically considered clerical staff and rarely allowed to advance to leading scientific roles or appear as authors on papers. In recent years, there has been a flurry of belated recognition of the contributions of human computers. A notable example is the film (and book) *Hidden Figures*, which tells the story of African-American women who worked as human computers for NASA and played an important role in the space race. The full story of human computing can be found in David Alan Gier's *When Computers Were Human*, [22].

Their roles as human computers led women to become the first professional computer programmers. Whatever arguments regarding the "rote" nature of the work that may have justified considering human computers unworthy of recognition or authorship clearly do not apply to computer programming. Implementing an algorithm in an efficient and reliable way requires substantial insight and creativity, even if the mathematical formulation of the algorithm is provided [24]. The transformation of the computer programmer from clerical staff to technically skilled engineer or scientist also coincided with the increase in recognition for the role and the general exclusion of women from it.

> Nowhere is this discounting of the women's contributions more evident than in the photographs that were used to publicize ENIAC. The first public photo appearing in the New York Times was a wide shot, showing both men and women working on the machine at various panels throughout the enormous computing room. Yet when the photograph was reprinted elsewhere, the women were usually cropped out. This was most obvious in an Army recruiting ad that appeared in various magazines later in 1946. Calling for "men with aptitude for scientific work" and extolling ENIAC as a prime example of "many amazing Army devices," the ad featured a heavily cropped version of the original Times photo, with only a single man shown working at a control panel. The message was clear: women need not apply; computers are for men. ... The disregard for women also signaled the beginning of the evolution of computer programming from a relatively unskilled clerical, "feminized" activity to a more technically skilled, supposedly more "masculine" pursuit. (Mark Wolverton, *Girl Computers* [24])

Nonetheless, women's roles at the first computer programmers meant that they performed foundational work in computational physics. The first programmers to conduct exploratory work on the MANIAC I (a successor to the ENIAC that was constructed at Los Alamos in 1952) were Mary Hunt and Mary Tsingou [38]. Mary Tsingou, a mathematician by training, went on to write the simulation of the Fermi–Pasta–Ulam–Tsingou (formerly Fermi–Pasta–Ulam) problem, which established that the existence of a small nonlinearity was *not* sufficient to guarantee equipartition of energy [38]. This was one of the first examples of a numerical experiment succeeding where no analytical method would suffice and producing an unexpected result. Despite her pivotal role, she was not listed as an author on the 1955 paper [39], and only in the last decade has it become common practice to include her name in the title of the problem [38].[19] Mary Tsingou also worked with J.R. Pasta to produce the first computer graphics [38] and later became one of the first experts in Fortran (the first high-level programming language) [38] and did more work on the FPUT problem [38, 40].[20]

At Los Alamos the urgent and secretive nature of the work and the isolated location caused it to become relatively common practice for scientists' wives (who were in many cases themselves skilled scientists and mathematicians) to be involved in technical work at the lab [25]. I will list a few examples here. Klara von Neumann (wife of John von Neumann) programmed the ENIAC and even modified the machine so it could run more difficult calculations [25] [31, p. 128] and later worked on MANIAC I. Augusta "Mici" Teller (wife of Edward Teller), a social scientist, performed calculations for the design of the atom bomb and later did preliminary programming work on MANIAC I for the Metropolis Algorithm [27].

Finally this brings us to Arianna Wright Rosenbluth. Born Arianna Wright [41, 42], she obtained a bachelor's degree from Rice University in 1946 [42] followed by an A.M. and Ph.D. in physics at Harvard[21] supervised by Van Vleck [41–44][22] and then became an Atomic Energy Commission postdoctoral fellow at Stanford University [27]. There she met Marshall Rosenbluth; the two moved to Los Alamos when Marshall was recruited to help with the hydrogen bomb [27]. She was fortunate that she *was* listed as an author on the critical "Metropolis Algorithm" paper [26], so her contribution was not completely erased. However, even her authorship did not guarantee recognition. A popular myth surrounding the Metropolis Algorithm claimed that it was invented at a cocktail party by the male authors of [26], and their wives were added as coauthors to reward them for enduring the boring technical conversation[23] [25]. The truth is, the first full computer implementation of Metropolis Monte Carlo was written *entirely* by Arianna Wright Rosenbluth [25, 27–29] at a time when programming a computer was a far more technically challenging and innovative task than it is today. This fact should be common knowledge among computational physicists, especially those

[19]The story of Mary Tsingou's contributions to physics is documented in [38].

[20]On this paper [40] Mary Tsingou was listed as an author under her married name: Mary Tsingou Menzel.

[21]Formally her degrees are from Radcliffe College [41]; at that time Harvard University did not admit women.

[22]Reference [43], Arianna Wright's dissertation, is available only in hard copy at the Harvard University library, but a condensed version was published in *Physical Review* [43].

[23]The fact that this apocryphal story takes place at a cocktail party completes the cliche of 1950s American sexism.

working in Monte Carlo, and sadly it is not. Young women deserve to see that computational physics is not merely a field where women can succeed, but a field that was in fact *founded* by women like Mary Tsingou Menzel and Arianna Wright Rosenbluth.

1.3 Condensed Matter Physics

This dissertation is in the field of *condensed matter physics*, which in the most informal sense possible, could be described as "the study of stuff that is not especially hot nor moving especially fast."[24] A more formal (but no less vague) definition is "the study of the behavior of large collections of interacting particles."[25] The haziness of this definition is appropriate since condensed matter is a very broad field encompassing the study of almost all everyday matter including liquids, solids, and gels as well as exotic matter like superconductors. Condensed matter physics is a tool for answering questions like: Why are some materials liquids? Why are others magnetic? What sorts of materials make good conductors of electricity? Why are ceramics brittle? Our understanding of condensed matter physics underlies much of modern technology; some prominent examples include ultra-precise atomic clocks, transistors,[26] lasers, and both the superconducting magnets and the superconducting magnetometers used for magnetic resonance imaging (MRI). Condensed matter physics overlaps with the fields of magnetism, optics, materials science, and solid-state physics.

Condensed matter physics is concerned with the behavior of large collections of particles. These particles are easy to define: they will sometimes be atoms or molecules and occasionally electrons and nuclei; condensed matter is almost never concerned with any behavior at higher energy scales (i.e., no need to worry about quarks). The key word in the definition is *large*. Atoms are very small, so any macroscopic amount of matter has a huge number of them, somewhere around Avogadro's number: 10^{23}. For practical purposes, we can assume any system that we study is infinite.[27] Large ensembles of particles display *emergent phenomena* that are not obvious consequences of underlying laws that govern the behavior of their microscopic components. In the words of Anderson [45]

> The ability to reduce everything to simple fundamental laws does not imply the ability to start from those laws and reconstruct the universe. ... hierarchy does not imply that science

[24]This definition distinguishes condensed matter from particle physics (the other broad subdiscipline of physics), which is the "study of really hot and really fast-moving objects."

[25]In practice, condensed matter tends to be the term used to describe physics that does not fit into one of the smaller, more well-defined subdisciplines like high-energy physics or cosmology.

[26]Both transistors and atomic clocks are essential to cellular telephones and satellite navigation systems like GPS.

[27]Hereafter we will also refer to infinite systems as "macroscopic" or as "the thermodynamic limit."

> X is "just applied Y." At each stage entirely new laws, concepts, and generalizations are necessary, requiring inspiration and creativity to just as great a degree as in the previous one. Psychology is not applied biology, nor is biology applied chemistry. (Anderson, *More is Different*)[28]

Emergent phenomena are not merely difficult to predict from the underlying microscopic laws, but they are effectively unrelated. At the most extreme scale, no one would argue that consciousness is somehow a property of standard model particles, or that democracy is a state that could ever be described in terms of quantum field theory. Here I will focus on two such emergent phenomena: phase transitions, where symmetries of the underlying laws are spontaneously violated and behavior is independent of microscopic details, and quasiparticles, an almost infinite variety of excitations of many-body states of matter that bear no resemblance to the "real" particles that make up the matter itself.

To highlight the importance of interactions, let us first consider the case of noninteracting particles. The canonical example here is the ideal gas, where the gas is composed of classical point-like particles that do not interact with each other. Because they do not interact, the motion of the particles is independent; if we want to know the energy of any particle, it is easy to calculate from its speed ($E = \frac{1}{2}mv^2$). The behavior of the whole system can be described by an ensemble of independent single particles. The partition function of an ensemble of N particles can be written as simply the product of the partition functions of independent individual particles.

$$Z_{\text{system}} = (Z_i)^N \tag{1.2}$$

This problem is separable [46]. Thermodynamic quantities like energy and specific heat can be extracted from this partition function. In the end, the behavior of the collection of particles is described by statistics of a single particle. As we will see, this is not the case for interacting systems.

When the particles are interacting things are very different. Instead of an ideal gas, let us consider a gas of classical electrons interacting via the Coulomb force $1/r$. For two electrons the equations of motion can be solved analytically, but in a solid there are 10^{23} electrons (for all practical purposes, we can round 10^{23} up to infinity). To write down the energy of one of them, we must account for the position of every single other electron. Thus the energy of just one electron is a function of $3N$ variables. Even with just three particles, analytic solutions are impossible in most cases. An analytic solution for the motion of 10^{23} electrons is impossible, and "it's not clear that such a solution, if it existed, would be useful" [47, p. 1]. This is many-body physics. Instead of following individual particles, we use tools

[28] This quote is taken from "More is different" by Anderson [45], an excellent refutation of reductionism and discussion of emergent phenomena written in a manner that should be accessible to non-physicists.

to describe their *collective motion* and the resulting emergent phenomena such as quasiparticles and phase transitions, which I will describe in the next few sections.

1.4 Classical Phase Transitions

One of the key emergent behaviors of large collections of interacting particles is the phenomenon of phase transitions. The most familiar phases of matter are solid, liquid, and gas and the most familiar phase transitions are the melting and boiling transitions between these phases. But phases of matter and the transitions between them come in a huge variety. Iron, for example, can be in a *paramagnetic* state (i.e., ordinary iron) or a *ferromagnetic* state (the one that sticks to your refrigerator). The ferromagnetic state can be destroyed by heating in much the same way as a solid can be melted.[29] Other examples of phases include different crystalline arrangements of a solid, plasmas, Bose–Einstein condensates, and superconductivity.

A phase transition is a qualitative change in the state of a system such as the onset of a net magnetization in the ferromagnetic transition or the onset of rigidity in the freezing transition. Phase transitions are associated with singularities in the free energy. The nature of this singularity can be used to classify the phase transition into one of two categories.[30] *First-order* phase transitions have a discontinuity in first derivative of the energy (i.e., the specific heat). First-order transitions are accompanied by the release or absorption of energy known as latent heat and at the critical point the two phases can coexist. The solid–liquid transition is an example of a first-order phase transition. After a few minutes, a glass of ice water will reach 0 °C, the only temperature at which both the ice and water can exist (at equilibrium). At 0 °C the specific heat of ice water is infinite, since adding energy to the system does not raise the temperature, it melts more ice; this situation only ends when all the ice has melted. The second category is known as *continuous* of phase transitions where there is a singularity in some higher-order derivative of the energy. As a system approaches a continuous phase transition, the *susceptibility* and correlation length diverge. Exactly at the critical point the correlations obey a power law and there are large fluctuations between the competing phases (there is no latent heat associated with a continuous phase transition). The destruction of the ferromagnetic state at high temperature is an example of a continuous phase transition.

In many cases phase transitions coincide with the spontaneous breaking of an underlying symmetry of the system and the formation of long-range order. For example, molecules in a liquid are packed closely together but without any long-range correlations; knowing the location of a molecule in one place does not provide

[29]The reader is not advised to try this at home. The ferromagnetic transition temperatures for most magnetic materials are well beyond the range of household ovens.

[30]There is a third category of "infinite-order" phase transitions such as the Berezinskii–Kosterlitz–Thouless (or BKT) transition. These will be discussed later in Sect. 4.1.3.

much information about the position of a distant molecule. This example highlights two key concepts in phase transitions: *correlations* and *spontaneous symmetry breaking*. As the liquid freezes, the molecules arrange themselves in a lattice to form an ordered structure. There are now long-range correlations between the locations of individual molecules. The solid has less symmetry than the liquid, so we say it has spontaneously broken translational symmetry. This symmetry breaking comes with long-range correlations and long-range order: the lattice means that distant molecules are now part of the same rigid structure. It might seem confusing to say that the solid has less symmetry; as an ordered lattice, it is highly symmetric, but the symmetries of the lattice as discrete rotations and translation, whereas the liquid has continuous translational and rotational symmetry, so we say the liquid had higher symmetry.

Both the singularity in the free energy and spontaneous symmetry breaking are emergent phenomena that occur only in infinite-size interacting systems [45]; finite-size systems cannot produce a singularity in the free energy

$$A = -k_B T \ln Z \tag{1.3}$$

because the free energy is a function of the partition function,

$$Z = \sum_i e^{-E_i/k_B T} \tag{1.4}$$

which is an analytic function of T for any finite-size system. Stationary states of finite-size systems are also prohibited from breaking an underlying symmetry of the Hamiltonian, but infinite-size systems get around this by having the characteristic time associated with statistical (or quantum) fluctuations to the competing symmetry-broken state diverge to be longer than the age of universe [45].

1.4.1 2D Ising Model

Rather than discuss phase transitions in the abstract, let us consider a concrete example: the 2D Ising ferromagnet on a square lattice. At each site there is a localized spin degree of freedom that can be either up or down $\sigma_i = \pm 1$ and interacts with its nearest neighbor like so

$$H = J \sum_{\langle i,j \rangle} \sigma_i \sigma_j \tag{1.5}$$

where $\langle i, j \rangle$ represents a sum over nearest neighbor pairs. For $J > 0$ the interactions are antiferromagnetic; for $J < 0$ they are ferromagnetic. The Ising model is considered to be a classical spin model; it has a sort of quantization in that $\sigma_i = \pm 1$, but there are no off-diagonal terms in the Hamiltonian so the system can be treated classically. In this sense it is a classical Hamiltonian acting on quantum spins.[31] Later we will discuss the quantum analog of the Ising model: the Heisenberg model. Despite its simplicity, the Ising model is rich in physics and it has the rare advantage that it can be solved exactly in both one and two dimensions. The Ising model was the first case where it was possible to prove that nonanalytic behavior can arise from a statistical mechanical system in the thermodynamic (infinite size) limit. The reader may want to consider visiting this site[32] where there is an interactive simulation of the 2D Ising model that displays Monte Carlo configurations in real time.

There are two degenerate minimum energy states of the Ising model: one has all the spins pointing up and the other has all spins pointing down. The Hamiltonian has perfect spin inversion (Z_2) symmetry. The natural state of a system is not the minimum of the energy, but the minimum of free energy,

$$A \equiv E - TS, \tag{1.6}$$

in which there is competition between minimizing energy and maximizing entropy. The two minimum energy states also correspond to minimum entropy states. The minimum energy states will therefore be unstable at any finite temperature. High temperatures will maximize the entropy, resulting in a roughly even mix of up and down and no net magnetization. As the temperature decreases, the system will want to minimize the energy by forming a net magnetization in one or the other direction. Will the entropy and symmetry prevail to maintain a net zero magnetization all the way to zero temperature? Or will energetic considerations dominate and symmetry be broken?

The Mermin–Wagner theorem [48] states that a discrete symmetry can be broken at finite temperatures in two dimensions, so we know there will be a finite-temperature phase transition. The 2D Ising model can also be solved exactly, so T_c is known exactly [49]:

$$\frac{kT_c}{J} = \frac{2}{\ln(1+\sqrt{2})} \approx 2.269 \tag{1.7}$$

For $T > T_c$ the system is disordered (there is no net magnetization) and for $T < T_c$ there is a net magnetization. This phase transition is continuous.

For our analysis, we will use the net magnetization, $\langle m \rangle$ as the order parameter. An order parameter captures the extent to which a system is in a given phase (which usually corresponds to some sort of long-range order). Order parameters are usually

---

[31]Permanent magnetism is in fact a fundamentally quantum phenomenon. A ferromagnetic transition occurs in the Ising model which matches the behavior of real materials, but the Ising-like degrees of freedom within those materials are the magnetic moments associated with electron *spin*, a fundamental property of electrons that has no classical interpretation.

[32]http://www.ibiblio.org/e-notes/Perc/ising.htm.

defined such that they are zero in the disordered phase and finite in the ordered phase. For example, in the gas–liquid transition, the order parameter is density (low in the gas phase and high in the liquid phase). In a disordered phase ($T > T_c$), the order parameter $\langle m \rangle$ is zero and there is no long-range order in the spin correlations, defined

$$C(r) \equiv \langle \sigma(0)\sigma(\vec{r}) \rangle \tag{1.8}$$

so they decay exponentially with correlation length ξ:

$$C(r) \propto e^{-r/\xi}. \tag{1.9}$$

This correlation length represents a physically important length scale in this system. At distances $r \gg \xi$ the correlations vanish, but at short distances $r \lessapprox \xi$, there are fluctuating pockets of order. Another way of thinking is that ξ represents the characteristic size of these pockets of order.

As the system approaches T_c from above, the distance from the critical point $t \equiv T - T_c$ becomes an important physical scale. This scale is related to the correlation length by a *critical exponent* ν [47, p. 231]. As the system approaches the $t = 0$ from above the correlation length diverges like

$$\xi \propto t^{-\nu}. \tag{1.10}$$

Fluctuations become extremely large and there are large pockets of order. Note that the order parameter (for an infinite system) will still be zero because the pockets of different competing orders will cancel each other out.

Below T_c, the Z_2 symmetry is spontaneously broken and $\langle m \rangle$ takes on some finite value $\langle m \rangle = \pm m(-t)$ that reflects the competition between spin alignment favored by energy and the fluctuations favored by entropy. For $T < T_c$ correlation length is again finite, but since there is long-range order, it now has the form of an exponential decay to a constant:

$$C(r) \propto e^{-r/\xi} + C_0. \tag{1.11}$$

The correlation length within the ordered phase describes typical size of fluctuations of the competing order.

Near the critical point, all thermodynamic quantities are governed by t through power laws. The susceptibility associated with ordering diverges as [47, p. 231]

$$\chi \propto t^{-\gamma}. \tag{1.12}$$

In our example of the 2D Ising model χ is the magnetic susceptibility. It diverges because the system is now "deciding" which direct to order in: up or down, an infinitesimal field will make the difference. The specific heat also diverges governed by the exponent α [47, p. 231]

$$C_v \propto t^{-\alpha}. \tag{1.13}$$

At exactly the critical point, $t = 0$, the correlation length becomes infinite. Here the system is not yet ordered, but is instead *critical*. The form of the correlation function is not exponential, but power law:

$$C(r) \propto r^{-(2-\eta)}. \tag{1.14}$$

Power laws are a form of scale-free behavior that appear at the critical point because the physically important scales in the system t and ξ have disappeared. Fluctuations of competing ordered phases appear at all sizes. This may seem abstract for magnetization, but it can be seen with the unaided eye in the phenomenon of *critical opalescence*, which occurs in binary mixtures of certain fluids which form a solution above some T_c and phase separate below T_c. At T_c, there are fluctuations between the mixed and unmixed phase at all length scales, including the wavelengths of visible light. As a result the mixture (which is otherwise clear) takes on a milky appearance around T_c.[33]

Near the critical point the dominant length scale of the system, ξ, is extremely large. As a result, microscopic details of the system (like the exact form of the interactions) should not matter. The critical exponents therefore depend only on the symmetry of the order parameter (i.e., the symmetry that is being broken) and the dimensionality of the system.[34] This principle is known as *universality*. Note that the phase boundaries themselves like T_c are not universal numbers and will depend on the microscopic details of the system. All liquid–gas transitions share the same symmetries and order parameter, so universality predicts that they should also share the same critical exponents describing the behavior around the critical point. To be clear, here we are discussing the liquid–gas transition along a critical isochore (line of constant density),[35] passing through the critical point so the transition is continuous.[36] Here there is no change in symmetry, but in the coexistence region there are two choices for the density high and low corresponding to liquid and gas, respectively [47, p. 166]. Indeed experiments show that the critical exponents for all substances that undergo this transition are the same even though the microscopic details of such systems vary greatly [50, p. 437]. More remarkably, those exponents also match the exponents of the ferromagnetic transition in the Ising model, which also has a scalar order parameter, but is otherwise almost completely different [50, p. 437]. In a very real sense the liquid–gas transition and the ferromagnetic transition are different instances of the same phenomenon, even though at first inspection they bear no resemblance to one another.

---

[33] See a time lapse of critical opalescence occurring here: https://youtu.be/DESZRUC8phw.

[34] Here we have assumed that the underlying interactions are short-range.

[35] The density can be fixed by enclosing particles in a fixed volume [47, p. 162].

[36] The everyday version of the liquid gas transition (boiling) occurs at constant pressure and is first order, not continuous [47, p. 162].

A consequence of universality is that there are a relatively limited number of universality classes into which all phase transitions fall. Once we know the critical exponents for *a* Z_2 phase transition in 2D, we know the critical exponents for *all* 2D Z_2 phase transitions. We can therefore build a "periodic table of phase transitions" with all the universality classes organized by symmetry and dimensionality. We can also choose to study the easiest realization of a universality class and that knowledge will apply rather generally to other transitions that fall in the same class. This becomes a key motivation for the theoretical and numerical study of both classical and quantum spin models. A spin model is in many cases the simplest and easiest to study realization of any universality class. The exact solution of the 2D Ising model [49] is a perfect example of this, but spin models like the Ising model and Heisenberg model are also exceptionally well-suited for numerical studies. Spin models are in this sense the "minimal models" for studying phase transitions. The subject of this dissertation, the J-Q model, is essentially a toy model invented to study the transition between the $O(3)$ Néel state and the Z_4 valence-bond solid (this will be discussed in more detail later).

Understanding phase transitions is critical to many technologies we use today. Air conditioning operates by using energy from a hot room to vaporize a working fluid, which is then compressed to release this energy outside. Nearly all electricity is generated using steam turbines, which rely on using a heat source boil water, producing steam to push the turbine. A promising future technology involves high-temperature superconductors. A superconductor is phase of matter that conducts electricity with zero resistance; developing superconductors that operate at or near room temperature would enable ultra-efficient power transmission, but requires an intimate understanding of (quantum) phase transitions.

1.5 Quantum Phase Transitions

In quantum condensed matter we study the properties of interacting matter that is governed by the laws of quantum mechanics. Intrinsically quantum effects—those that cannot be described by some effective classical description—tend to appear at extremely low temperatures. The classical interacting electron gas that we discussed before was already impossible to study in terms of the motion of individual particles; the quantum electron gas is harder still. Electrons are not really billiard balls with well-defined positions and momenta—they are quantum objects described by a wavefunction. Further, electrons are identical particles, so in no sense can we say that we will follow the motion of any particular electron (we cannot "paint one red"). Once we place an electron into the material it ceases to exist as an independent particle, it is now mixed up with every other electron. *The electron is gone.* Instead the whole system occupies a *many-body* state

$$\psi(\vec{r}_1, \vec{r}_2, \ldots) \tag{1.15}$$

which is a function of the positions of every electron in the material.

Fortunately, we almost never have to consider the full quantum many-body state. We are concerned primarily with the low temperature (and therefore low energy) properties of these materials, so most systems can be described in terms of relatively few effective degrees of freedom. For example, in almost all cases only the outer (valence) shell of electrons will interact with other atoms. Inner-shell electrons are strongly bound to the nucleus and can be absorbed into an effective nuclear potential. Even with the elimination of inner-shell electrons the remaining problem is still a difficult many-body physics problem. In the case of insulating materials, where the valence electrons are tightly bound, we can make the additional simplification of treating the electrons as occupying discrete orbitals on each site and treating the overlap of these orbitals as a "hopping" between the sites. Under some circumstances these models can then be transformed into a model of localized electrons with short-range spin–spin interactions: quantum spin models.

We here will consider a class of models known as quantum spin models. These represent solids as a lattice of localized spin degrees of freedom. The simplest of these spins models is the $S = \frac{1}{2}$ (quantum) Heisenberg model, which consists of a lattice of sites each hosting $S = \frac{1}{2}$ that interact with their nearest neighbors via a Hamiltonian given by:

$$H = J \sum_{\langle i,j \rangle} \mathbf{S}_i \cdot \mathbf{S}_j, \tag{1.16}$$

where $\langle i, j \rangle$ denotes a sum over nearest neighbors on some lattice, $J > 0$ is the antiferromagnetic (AFM) case, and $J < 0$ is the ferromagnetic (FM) case. This represents a dramatic simplification of a material where the Hilbert space would consist of the positions of all electrons (at least a few per atom) in three-dimensional space, to a system composed of a lattice of discrete sites which can only occupy two states: spin up or spin down. Even with these simplifications, the Heisenberg model remains a difficult interacting many-body problem. In the S^z basis, the Hilbert space is all possible combinations of the S^z components of each site, for example, $|\uparrow\uparrow\downarrow\uparrow\rangle$. Therefore the state space is exponentially large: 2^N.

Let us compare the 2D Heisenberg antiferromagnet (AFM) to the Ising antiferromagnet, both on a square lattice. The order parameter of the Ising model is the staggered magnetization,

$$\langle m_s \rangle = \sum_{x,y} (-1)^{(x+y)} \sigma(x, y), \tag{1.17}$$

a *scalar*. In the Heisenberg model, however, the spins are *vectors* and the order parameter, the staggered spin polarization,

$$\langle \vec{S}_s \rangle = \sum_{x,y} (-1)^{(x+y)} \vec{S}(x, y), \tag{1.18}$$

is a vector as well. The ordered state in the Ising model breaks a discrete symmetry, so it can occur at finite temperature in two dimensions, but the Heisenberg model has *continuous* ($O(3)$) symmetry, which in 2D can only be spontaneously broken at zero temperature.[37] The ground state of the Ising model is the Néel state, a checkerboard pattern of alternating up and down spins. The Heisenberg model is also antiferromagnetic: it wants to align neighboring spins in opposite directions. Unlike the Ising model, the Néel state is *not* an eigenstate of the Heisenberg model.[38] This can be more clearly seen by rewriting the interaction $\vec{S}_i \cdot \vec{S}_j$ in terms of S^z, S^+, and S^-:

$$H = J \sum_{\langle i,j \rangle} \left[S_i^z S_j^z + \frac{1}{2} \left(S_i^+ S_j^- + S_i^- S_j^+ \right) \right]. \tag{1.19}$$

From this representation it is clear that when this Hamiltonian acts on the Néel state it will change it. The ground state therefore cannot be the Néel state. Instead, the ground state can be described as a Néel state oriented along some symmetry-broken polarization axis and dressed by *quantum fluctuations*.

Unlike the thermal fluctuations present in the Ising model, quantum fluctuations persist all the way down to zero temperature and can drive something impossible in classical systems: zero-temperature continuous phase transitions—quantum phase transitions. Quantum phase transitions strictly occur at absolute zero, and correspond to a level crossing between the ground state and the first excited state which produces a totally new ground state.[39] Many of the tools developed to understand classical phase transitions can be applied directly to quantum phase transitions. Under most circumstances, quantum effects and quantum fluctuations occur only at very small scales. At a quantum critical point, instead of macroscopic thermal fluctuations there are *macroscopic quantum fluctuations*, which are interesting to study both for fundamental physics and possible device applications.

To study states with strong quantum fluctuations, we usually need to introduce competing interactions. One of the simplest ways to do this is to add a competing next-nearest neighbor interaction (the J_1-J_2 model [51, 52]):

$$H = J_1 \sum_{\langle i,j \rangle} \mathbf{S}_i \cdot \mathbf{S}_j + J_2 \sum_{\langle\langle i,j \rangle\rangle} \mathbf{S}_i \cdot \mathbf{S}_j \tag{1.20}$$

[37] This restriction is set by the Mermin–Wagner Theorem [48].

[38] In the ferromagnetic case, the ground state of the Heisenberg model is simply the fully polarized state, which *is* an eigenstate of the Hamiltonian and therefore there are no quantum fluctuations in the ground state of the Heisenberg ferromagnet. The Heisenberg antiferromagnet is more interesting than the ferromagnet because it exhibits stronger quantum fluctuations.

[39] At finite size, this is an "avoided level crossing" where the two states hybridize in some way, but in the thermodynamic limit they cross.

where $\langle i, j\rangle$ represents a sum over nearest neighbors and $\langle\langle i, j\rangle\rangle$ represents a sum over next-nearest neighbors. If both interactions are antiferromagnetic, then this Hamiltonian is frustrated: there is no classical arrangement of up and down spins that satisfies all interactions. Frustration tends to produce strong quantum fluctuations, precisely the situation that we want to study. One of the interesting quantum states that occurs in the J_1-J_2 model is the valence-bond solid (VBS). This state will play an important role in this thesis, and is discussed extensively in the following chapters, especially Chap. 4, but we will describe it briefly here. In the VBS, sites pair up with their neighbors to form an ordered arrangement of singlet bonds (see Fig. 4.1). This breaks discrete lattice symmetry (Z_4), but respects $O(3)$ spin-rotation symmetry (which is violated by the Néel state). This state tends to occur in frustrated systems like the J_1-J_2 model. Frustrated spin models almost always suffer from the *sign problem*, which makes them inaccessible to quantum Monte Carlo and therefore inaccessible to large-scale simulations.

The solution to this problem is a toy model with a strange-looking interaction—the J-Q model. The J-Q model is a numerical method in its own right; it augments the AFM Heisenberg model with a competing four-spin interaction

$$H = J\sum_{\langle i,j\rangle} \vec{S}_i \cdot \vec{S}_j - Q \sum_{\langle i,j,k,l\rangle} \left(\frac{1}{4} - \vec{S}_i \cdot \vec{S}_j\right)\left(\frac{1}{4} - \vec{S}_k \cdot \vec{S}_l\right) \tag{1.21}$$

Here $\langle i, j\rangle$ still represents a sum over nearest neighbors and $\langle i, j, k, l\rangle$ represents a sum over four spins in a row $(i, i+1, i+2, i+3)$ in 1D and over plaquettes with "bonds" i, j and k, l arranged as parallel links in the horizontal $\begin{smallmatrix} k & l \\ i & j \end{smallmatrix}$ and vertical $\begin{smallmatrix} j & l \\ i & k \end{smallmatrix}$ directions on the 2D square lattice. Using singlet projection operators defined,

$$P_{i,j} \equiv \frac{1}{4} - \vec{S}_i \cdot \vec{S}_j, \tag{1.22}$$

and adding a constant energy offset, this Hamiltonian can be written more compactly:

$$H = -J\sum_{\langle i,j\rangle} P_{i,j} - Q \sum_{\langle i,j,k,l\rangle} P_{i,j}P_{k,l}. \tag{1.23}$$

The J-Q model is sign-problem free on bipartite lattices when $J, Q \geq 0$. The Q term therefore provides a stand-in for conventional frustration allowing the exploration of behavior that usually occurs in frustrated systems. One example of this behavior is metamagnetism, a first-order phase transition associated with a discontinuity in the magnetization (see Chaps. 2 and 3) [11, 12]. The Q term also drives a quantum phase transition from the Néel state to a valence-bond solid (VBS), and example of deconfined quantum criticality. This phase transition has been well-studied in the literature [1, 4, 5, 9, 10, 17–19] and will be described in more detail in Chap. 4, so here I will offer only brief comments to provide motivation.

1.5.1 Deconfined Quantum Criticality

There is substantial numerical evidence that the transition between the Néel state and the valence-bond solid in the 2D J-Q model is both direct (i.e., there is no intermediate phase) and continuous [1, 4, 5, 9, 10, 17–19]. This is a violation of the Landau–Ginzburg paradigm which predicts that (absent fine tuning) direct phase transitions between phases breaking unrelated symmetries (like $O(3)$ in the Néel state and Z_4 in the VBS) should be first order [16]. In the Landau–Ginzburg paradigm critical points are described by the order parameter that appears in the ordered phase. The order parameter of the Néel state is the Néel polarization vector and the corresponding Goldstone modes are spin waves—gapless bosonic magnons carrying $S = 1$. In the VBS the excitations are triplons—triplet waves formed by breaking one of the singlet bonds (see Fig. 4.1). These triplons are gapped and carry $S = 1$.

The solution to the violation of the Landau–Ginzburg paradigm is deconfined quantum criticality (DQC) [15, 16]. Instead of being described by the excitations or order parameters of either ordered phase, the critical point is instead described by emergent fractionalized excitations. In this case these excitations are known as spinons and are bosons carrying $S = \frac{1}{2}$ (the lattice breaks Lorenz symmetry, so the Spin-Statistics Theorem does not apply and half-integer spin particles need not be fermions). In either ordered phase, the spinons are confined within the elementary excitations of that phase: magnons in the Néel phase and triplons in the VBS. This confinement is similar to the confinement of quarks in a proton. The spinons are only deconfined at the critical point. This phenomenon will be described in more detail in Sect. 4.1.1.

Here it is worthwhile to point out that in most cases, quantum fluctuations can be eliminated through a clever choice of basis. For example, we can construct a Hamiltonian for which the ground state is an exact VBS, and if we write that state in its natural basis, then the ground state is simply a product state of singlets on all the bonds. What is unique about quantum critical points is that these quantum fluctuations will exist in *any* basis and the scale of these fluctuations will be divergent, leading to macroscopic quantum effects.

1.5.2 What Are Quasiparticles?

In this dissertation I will discuss a variety of quasiparticles from relatively mundane magnons to exotic spin-half bosons. I thought it was worthwhile here to include a few remarks on what quasiparticles mean.[40] As was discussed earlier, studying many-body systems by following the motion each constituent particle is a hopeless

[40]There is a surprisingly cogent discussion of this topic in the Wikipedia article "Quasiparticle" which can be accessed at the following https://en.wikipedia.org/wiki/Quasiparticle.

endeavor. Instead, we consider the many-body quantum ground state as a vacuum, and consider excitations on that vacuum.[41] These excitations constitute emergent particles—quasiparticles—which more naturally describe the behavior of the system. Quasiparticles encapsulate the collective motion of the underlying many-body system. In a sense this is keeping track of what is *happening* rather than what is there. Usually these can be formulated such that their density is relatively low and they are weakly interacting or noninteracting. The immediate benefit of this is obvious: fewer things to keep track of, but it is less obvious what these excitations mean and in what sense we should think of them as "real" particles.

A good example of a quasiparticle (and its utility) is a phonon. Consider striking an aluminum xylophone bar with a mallet. The aluminum bar is an orderly lattice of aluminum atoms. When the mallet hits the bar the electrons in the mallet are definitely interacting with the electrons in the bar, but thinking of this even in terms of electrons is an intractable problem. Instead, the mallet strike excites *phonons*. What is a phonon? It is a lattice vibration quantum, a sound wave in the lattice of the aluminum bar. We cannot pull a phonon out of the bar to study it in isolation. Phonons only exist inside the bar; they only make sense in the context of the vacuum of which they are an excitation. A phonon itself is a massless boson that bears no resemblance to any of the ingredients used to make the bar (electrons, protons, and neutrons) which are all massive fermions. You might say then that these are not "real" particles, and to some extent you would be right, phonons are *quasiparticles*: excitations of the many-body state.[42]

In some cases the quasiparticle excitations of a solid bear a striking resemblance to the original electrons. In these materials, the excitations are electron-like quasiparticles (carrying charge $-e$ and $S = \frac{1}{2}$) and the interactions with the other electrons can be absorbed to a "renormalized" mass. Such materials are sometimes referred to as "single-electron" materials, but they are in reality many-body systems that happen to have electron-like excitations.[43] Even in so-called single-electron materials there are other non-electron quasiparticles, like phonons and holes. A hole is an unoccupied state (below the Fermi surface) where an electron could be. Holes are antielectrons, they carry $S = \frac{1}{2}$ and the opposite charge, $+e$. An incoming photon can excite an electron-hole pair which can later meet and annihilate, releasing the energy used to create the pair (although in a solid this energy is not the same as their inertial mass).[44] In fact, this is more

[41]This approach works in classical physics as well, but since we are concerned with quantum many-body physics we will describe these in quantum language.

[42]Connecting to the previous discussion about spontaneous symmetry breaking, phonons are the massless Goldstone bosons that arise from breaking a continuous symmetry, in this case the translational symmetry that was broken to form the lattice.

[43]These "single electron" materials are typically materials where the electrons interact with long-range Coulomb-like interactions, so the Coulomb potential is averaged over many distant electrons and varies slowly in space. Ironically, when the interactions are short-range, materials are harder to describe in terms of single-electron physics.

[44]This process is part of how solar panels work.

than a metaphor. Electron-position pairs are excitations of the "real" vacuum in the exact same way that electron-hole pairs are excitations of the vacuum of a solid. In p-type semiconductors the charge carriers are in fact holes; this can even be confirmed with Hall effect experiments. We have now seen that the quasiparticle excitations of a solid need not bear any resemblance to the "ingredients" used to make the solid. In fact, the variety of quasiparticles that can be created in solids is far greater than the variety of "real" particles that exist in free space (the "real" vacuum) because there is only one "real" vacuum, but infinite variations on the vacuum that is present in solids. By doping, tweaking the arrangement of the atoms, or adding external fields we can engineer a huge variety of vacuums and therefore a huge variety of quasiparticles (a few examples include plasmons, polarons, and magnons). We can manually break symmetries, tune coupling constants, or even change the dimensionality[45] to create particles that cannot exist in free space. A great example of this are the spinons that occur at the Néel-VBS transition ($S = \frac{1}{2}$ bosons). In three-dimensional free space, the Spin-Statistics Theorem (which relies on Lorenz symmetry) requires that all bosons have integer spin, but there is no Lorenz symmetry in a solid and therefore no such restriction.

Here I would like to make two remarks about such exotic quasiparticles. The first is that they are real. Quasiparticles exist as excitations of a vacuum, but *all* particles exist as excitations of a vacuum, including the Standard Model particles that are typically considered to be fundamental. The only difference between quasiparticles and "real" particles is that for quasiparticles, we know what the vacuum is, whereas Standard Model particles are excitations of the vacuum we live in. We have no reason to believe that the vacuum we live in is somehow more fundamental than any other vacuum. The second remark is that these exotic quasiparticles are emergent phenomena. Effectively, spinons are a type of matter that can exist, and the fact that they are made out of protons, neutrons, and electrons in this case does not make spinons a property of protons, neutrons, and electrons any more than the plot of a novel is a property of the protons, neutrons and electrons that make up the pages and ink. Not only is the existence of spinons not obvious from the underlying laws, but it is in a sense unrelated to those laws. Indeed, "more is different" [45].

1.6 Motivation

I will now briefly describe some motivation for pursuing this work starting from the most broad reasons concerning basic research and continuing to address the specific reasons for the work conducted here. Scientific research falls on a spectrum from the most applied (engineering a crumple zone to protect occupants of a car) to the most basic research (like the ongoing searches for dark matter). On that spectrum the

[45] Admittedly, experiments are currently limited to reducing the dimensionality (to quasi-2D, quasi-1D, and quasi-0D).

work described here falls closer to the latter example. One might ask: why perform basic research with no obvious application instead of applied research? The key word here is *obvious*; applied research is really how one describes research where the application is obvious. Applied research stands on a foundation of decades of basic research. Designing the Global Positioning System is applied research, but developing the theory of general relativity that is required to make that system work is basic research. When Einstein was formulating general relativity, there were no obvious applications, and even a genius could not have foreseen the eventual application to accurate satellite-based positions that required a myriad of other not-yet-invented technologies such as extremely accurate atomic clocks and transistors. Basic research describes scientific research where the applications are *not* obvious, and if they exist it may be decades before they become apparent. In pursuing basic research we are making an investment in the long-term well-being of our society, laying the foundation for future technologies, medicines, and understanding.

Now, to be much more specific: why study spin models? The first answer is that these models are often accurate descriptions of real materials and many fundamental technologies rely on magnetic materials, from the strong rare-earth permanent magnets that make it possible to build tiny electric motors to power spinning hard disk drives to the giant magnetoresistance that enables reliable high-density magnetic storage on those same drives. There are also many materials that do not display net magnetization, but are well described by localized magnetic moments with short-range (usually antiferromagnetic) interactions. For example, I am currently engaged in a collaboration with the group of Arthur Ramirez at University of California, Santa Cruz studying materials which behave as three-dimensional arrays of coupled spin chains (with antiferromagnetic Heisenberg-like interactions that are strong within the chains and weak between them) in the presence of an external field [53] where I am using my simulations to directly compare to their experiments.

The transistor-fueled digital computing revolution that has powered everything from the space race to smartphones to the numerical tools used in this thesis and the laptop that I am currently writing it on was powered by an underlying understanding of the single-electron physics of the solid-state devices. The next generation of devices will rely on the burgeoning understanding of many-body physics. Spin systems have historically been critical to our understanding of classical phase transitions, and are now the foundation of our understanding of quantum phase transitions. A key application of quantum phase transitions is in the understanding of high-T_c superconductivity, a more detailed understanding of which could enable room-temperature superconductors that would revolutionize computing, energy, and transportation. Quantum spin systems will undoubtedly play a role in developing our understanding of basic physics, and may also form the basis of critical new technologies such as quantum computers. An understanding of these phase transitions may even have applications in cosmology and our basic understanding of the universe [54].

References

1. A.W. Sandvik, Phys. Rev. Lett. **98**, 227202 (2007). https://doi.org/10.1103/PhysRevLett.98.227202
2. A.W. Sandvik, in *American Institute of Physics Conference Series*, vol. 1297, ed. by A. Avella, F. Mancini (2010), pp. 135–338. https://doi.org/http://arxiv.org/abs/1101.3281
3. R.G. Melko, R.K. Kaul, Phys. Rev. Lett. **100**, 017203 (2008). https://doi.org/10.1103/PhysRevLett.100.017203
4. A.W. Sandvik, Phys. Rev. Lett. **104**, 177201 (2010). https://doi.org/10.1103/PhysRevLett.104.177201
5. A.W. Sandvik, V.N. Kotov, O.P. Sushkov, Phys. Rev. Lett. **106**, 207203 (2011). https://doi.org/10.1103/PhysRevLett.106.207203
6. S. Sanyal, A. Banerjee, K. Damle, Phys. Rev. B **84**, 235129 (2011). https://doi.org/10.1103/PhysRevB.84.235129
7. Y. Tang, A.W. Sandvik, Phys. Rev. Lett. **107**, 157201 (2011). https://doi.org/10.1103/PhysRevLett.107.157201
8. Y. Tang A.W. Sandvik, Phys. Rev. B **92**, 184425 (2015). https://doi.org/10.1103/PhysRevB.92.184425
9. H. Suwa, A. Sen, A.W. Sandvik, Phys. Rev. B **94**, 144416 (2016). https://doi.org/10.1103/PhysRevB.94.144416
10. H. Shao, W. Guo, A.W. Sandvik, Science **352**, 213 (2016). https://doi.org/10.1126/science.aad5007
11. A. Iaizzi, A.W. Sandvik, J. Phys. Conf. Ser. **640**, 012043 (2015). http://stacks.iop.org/1742-6596/640/i=1/a=012043
12. A. Iaizzi, K. Damle, A.W. Sandvik, Phys. Rev. B **95**, 174436 (2017). https://doi.org/10.1103/PhysRevB.95.174436
13. S. Sachdev, T. Senthil, R. Shankar, Phys. Rev. B **50**, 258 (1994). https://doi.org/10.1103/PhysRevB.50.258
14. A. Iaizzi, K. Damle, A.W. Sandvik, Phys. Rev. B **98**, 064405 (2018). https://doi.org/10.1103/PhysRevB.98.064405
15. T. Senthil, A. Vishwanath, L. Balents, S. Sachdev, M.P.A. Fisher, Science **303**, 1490 (2004). https://doi.org/10.1126/science.1091806
16. T. Senthil, L. Balents, S. Sachdev, A. Vishwanath, M.P.A. Fisher, Phys. Rev. B **70**, 144407 (2004). https://doi.org/10.1103/PhysRevB.70.144407
17. J. Lou, A.W. Sandvik, N. Kawashima, Phys. Rev. B **80**, 180414 (2009). https://doi.org/10.1103/PhysRevB.80.180414
18. S. Jin, A.W. Sandvik, Phys. Rev. B **87**, 180404 (2013). https://doi.org/10.1103/PhysRevB.87.180404
19. Y. Tang, A.W. Sandvik, Phys. Rev. Lett. **110**, 217213 (2013). https://doi.org/10.1103/PhysRevLett.110.217213
20. A. Iaizzi, H.D. Scammell, O.P. Sushkov, A.W. Sandvik, Direct numerical observation of Bose-Einstein Condensation of deconfined spinons (2018) (in preparation)
21. S. Brandt, *Data Analysis: Statistical and Computational Methods for Scientists and Engineers*, 3rd edn. (Springer, Berlin, 1998)
22. D. Grier, *When Computers Were Human* (Princeton University Press, Princeton, 2013). https://books.google.com/books?id=YTcDAQAAQBAJ
23. M. Campbell-Kelly, Sci. Am. **301**, 62 (2009). https://doi.org/10.1038/scientificamerican0909-62
24. M. Wolverton, American Heritage **61**, 201 (2011). Girl Comput. http://www.americanheritage.com/content/girl-computers
25. J.E. Gubernatis, AIP Conf. Proc. **690**, 3 (2003). https://doi.org/10.1063/1.1632111
26. N. Metropolis, A.W. Rosenbluth, M.N. Rosenbluth, A.H. Teller, E. Teller, J. Chem. Phys. **21**, 1087 (1953). https://doi.org/10.1063/1.1699114

27. J.E. Gubernatis, Phys. Plasmas **12**, 057303 (2005). https://doi.org/10.1063/1.1887186
28. K.-H. Barth, Interview of Marshall Rosenbluth, Niels Bohr Library and Archives. Published by American Institute of Physics (2003). https://www.aip.org/history-programs/niels-bohr-library/oral-histories/28636-1
29. M.N. Rosenbluth, AIP Conf. Proc. **690**, 22 (2003). https://doi.org/10.1063/1.1632112
30. W.K. Hastings, Biometrika **97** (1970). https://doi.org/10.1093/biomet/57.1.97
31. N. Metropolis, Los Alamos Sci. **15**, 125 (1987). Available online at https://library.lanl.gov/cgi-bin/getfile?00326866.pdf
32. R. Ivie, K.N. Ray, AIP Rep. (2005). Women Phys. Astron. https://www.aip.org/statistics/reports/women-physics-and-astronomy-2005
33. Fraction of bachelor's degrees earned by women, by major . https://www.aps.org/programs/education/statistics/womenmajors.cfm. Credit: American Physical Society, Source: IPEDS Completion Survey. Accessed 3 Nov 2017
34. Doctoral degrees earned by women, by majorhttps://www.aps.org/programs/education/statistics/fraction-phd.cfm, Credit: American Physical Society, Source: IPEDS Completion Survey. Accessed 3 Nov 2017
35. R. Ivie, S. White, R.Y. Chu, Phys. Rev. Phys. Educ. Res. **12**, 020109 (2016). https://doi.org/10.1103/PhysRevPhysEducRes.12.020109
36. G. Ghiasi, V. Larivière, C.R. Sugimoto, PLoS One **10**, 1 (2016). https://doi.org/10.1371/journal.pone.0145931
37. N. Caplar, S. Tacchella, S. Birrer, Nat. Astron. **1**, S41550 (2017). https://doi.org/10.1038/s41550-017-0141
38. T. Dauxois, Phys. Today **61**, 55 (2008). https://doi.org/10.1063/1.2835154
39. E. Fermi, P. Pasta, S. Ulam, M. Tsingou, Los Alamos Rep. (1955). https://doi.org/10.2172/4376203. Author's note: although Tsingou is not listed as an author on the original paper, the preferred citations of the paper has since been amended to include Tsingou. https://www.osti.gov/biblio/4376203-studies-nonlinear-problems
40. J. Tuck, M. Menzel, Adv. Math. **9**, 399 (1972). https://doi.org/10.1016/0001-8708(72)90024-2
41. Harvard University Alumni Directory https://alumni.harvard.edu/help/directory-search (2017). Accessed 20 Nov 2017. Not accessible to public
42. Harvard Physics PhD Theses 1873–1953 Harvard Physics Department Website: https://www.physics.harvard.edu/uploads/files/thesesPDF/PhD1873-1953.pdf (2017). Accessed 21 Nov 2017
43. A. Wright, Phys. Rev. **76**, 1826 (1949). https://doi.org/10.1103/PhysRev.76.1826
44. A. Wright, Some Aspects of Paramagnetic Relaxation. Ph.D. thesis, Radcliffe College (1949). Supervised by J.H. Van Vleck at Harvard University
45. P.W. Anderson, Science **177**, 393 (1972). https://doi.org/10.1126/science.177.4047.393
46. J. Hodgman, *Vacationland: True Stories from Painful Beaches* (Penguin Publishing Group, London, 2017). https://books.google.com/books?id=7BgcDgAAQBAJ
47. P.M. Chaikin T.C. Lubensky, *Principles of Condensed Matter Physics* (Cambridge University Press, Cambridge, 1998)
48. N.D. Mermin, H. Wagner, Phys. Rev. Lett. **17**, 1133 (1966). https://doi.org/10.1103/PhysRevLett.17.1133
49. L. Onsager, Phys. Rev. **65**, 117 (1944). https://doi.org/10.1103/PhysRev.65.117
50. R.K. Pathria, P.D. Beale, *Statistical Mechanics*, 3rd edn. (Elsevier, Amsterdam, 2011)
51. C.K. Majumdar, D.K. Ghosh, J. Math. Phys. **10**, 1388 (1969). https://doi.org/10.1063/1.1664978
52. C.K. Majumdar, D.K. Ghosh, J. Math. Phys. **10**, 1399 (1969). https://doi.org/10.1063/1.1664979
53. N. Blanc, J. Trinh, L. Dong, X. Bai, A.A. Aczel, M. Mourigal, L. Balents, T. Siegrist, A.P. Ramirez, Nat. Phys. **14**, 273 (2017). https://doi.org/10.1038/s41567-017-0010-y
54. R.B. Laughlin, Science **303**, 1475 (2004). https://doi.org/10.1126/science.1095266

Chapter 2
Saturation Transition in the 1D J-Q Model

2.1 Introduction

In this chapter we characterize the magnetization process of a one-dimensional Heisenberg antiferromagnet with four-spin interactions of strength Q in addition to the standard antiferromagnetic exchange term of strength J (the J-Q model [2, 3]) as it is subjected to an external magnetic (Zeeman) field. The model is defined in terms of singlet projectors acting on a lattice of $S = 1/2$ sites:

$$P_{i,j} \equiv \frac{1}{4} - \mathbf{S}_i \cdot \mathbf{S}_j. \tag{2.1}$$

The standard antiferromagnetic Heisenberg exchange is equivalent to $-JP_{ij}$ with $J > 0$. In the J-Q model this interaction is supplemented by the product $-QP_{i,j}P_{k,l}$ (or products of more than two projectors [4]) with the site pairs i, j and k, l suitably arranged and summed over the lattice sites with all lattice symmetries respected. The long-range ordered (in two or three dimensions) or critical (in one dimension) antiferromagnetic (AFM) state of the pure Heisenberg model can be destroyed for sufficiently large Q/J. A non-magnetic ground state with broken lattice symmetries due to dimerization (a valence-bond solid, VBS) then appears. The VBS state and the quantum phase transition between the AFM and VBS states have been studied extensively in both one [5–7] and two [2, 8–11] dimensions. The

This chapter is a lightly edited version of a paper, "Field-driven quantum phase transition in $S = \frac{1}{2}$ spin chains" coauthored with Anders W. Sandvik and Kedar Damle appearing in [1]. Reprinted with permission.

A. Iaizzi, *Magnetic Field Effects in Low-Dimensional Quantum Magnets*, Springer Theses, https://doi.org/10.1007/978-3-030-01803-0_2

J-Q model is a member of a broad family [4] of Marshall positive spin Hamiltonians constructed from products of any number of singlet projection and permutation operators.

Here we consider the simplest one-dimensional (1D) J-Q model, where the Q term is composed of a product of just two singlet projection operators:

$$H_{JQ} = -J \sum_i P_{i,i+1} - Q \sum_i P_{i,i+1} P_{i+2,i+3}, \tag{2.2}$$

and add an external magnetic field of strength h_z to define the J-Q-h model:

$$H_{JQh} = H_{JQ} - h_z \sum_i S_i^z. \tag{2.3}$$

We set the energy scale by fixing $J = 1$ and refer to the dimensionless parameters $q \equiv Q/J$ and $h \equiv h_z/J$.

Our focus will be on the magnetization curve as a function of the field, which we study both at $T = 0$ and $T > 0$. We use the stochastic series expansion (SSE) [3, 12] quantum Monte Carlo (QMC) method with directed loop updates [13], supplemented by quantum replica exchange [14, 15] to alleviate metastability problems in the simulations. We show that the Q term has dramatic consequences for the magnetization process. In the pure Heisenberg chain ($q = 0$), and for small q, the magnetization curve at temperature $T = 0$ is continuous. When q exceeds a critical value, a magnetization jump (metamagnetic transition) [16, 17] appears between a partially magnetized and the fully polarized state. Using an ansatz motivated by numerical results for two magnons in a saturated background, we obtain an exact analytical result for the minimum coupling ratio, $q_{\min}$, at which such a magnetization jump can occur; $q_{\min} = \frac{2}{9}$. This calculation also reveals the mechanism of the magnetization jump: the onset of attractive magnon interactions when $q > q_{\min}$. At exactly $q_{\min}$, the magnons behave as effectively noninteracting particles. The onset of a bound state of magnons is a general mechanism for metamagnetism [18, 19], but normally this phenomenon has been associated with frustration due to competing exchange couplings [18–25] or strong spin anisotropy [18, 20, 21] (including the classical two-dimensional (2D) Ising model with second-neighbor interactions [26, 27]). We believe this effect could also explain the metamagnetic transition reported in a ring exchange model [28], (a close relative of the J-Q model), where the metamagnetic transition corresponds to a first-order transition from a partially occupied to a fully occupied state. Our study provides an example of metamagnetism in a spin-isotropic system without traditional frustration. Note that the onset value $q_{\min} = \frac{2}{9}$ of metamagnetism is much smaller than the critical value $q_c \approx 0.85$ at which the chain dimerizes in the absence of a field. Thus, the metamagnetism here is not directly related to the VBS state of the J-Q model.

A bound state of magnons *does not* occur in the standard J_1-J_2 Heisenberg chain [29–31] with frustrated antiferromagnetic couplings $J_1 > 0$, $J_2 > 0$, but it *does* occur [19, 22, 24] for the also-frustrated FM-AFM regime $J_1 < 0$, $J_2 > 0$. In our study of the unfrustrated regime, we find bound magnon states in the J_1-J_2 chain with a ferromagnetic (FM) second-neighbor coupling (AFM $J_1 > 0$, FM $J_2 < 0$), but only if this second-neighbor coupling is also spin anisotropic, of the form $J_2[S_i^z S_j^z + \Delta(S_i^x S_j^x + S_i^y S_j^y)]$. The existence of a bound state for some values of the parameters $\Delta \neq 0$ and $|J_2/J_1|$ is likely a precursor to a metamagnetic transition as in the J-Q-h chain, but we do not study it further with QMC here.

We also study the J-Q-h chain at $T > 0$ in the region close to magnetic saturation when $q \leq q_{\min}$. Here one would expect the dependence of the magnetization on the field and the temperature to be governed by a remarkably simple "zero-scale-factor" universal critical scaling form [32]. We observe this behavior clearly for $q = 0$ and $q \ll q_{\min}$. For q closer to $q_{\min}$ we find that the scaling form is only obeyed at extremely low temperatures, due to onset of metamagnetism at $q = q_{\min}$. We expect $q_{\min}$ to be a tricritical point at which the sign of the quartic coupling ($|\psi|^4$) of the boson field changes in the low-energy effective field theory of the system. This corresponds to the two-magnon interaction switching from repulsive to attractive at this point. Precisely at $q = q_{\min}$, the two-magnon interaction vanishes and the system is dominated by three-body interactions, represented in the effective field theory by a $|\psi|^6$ term which is marginal in $d = 1$. The smallness of the quartic term close to $q_{\min}$ leads to a cross-over, which we observe, between tricritical and zero-scale-factor behavior, with the cross-over temperature approaching zero as $q \to q_{\min}$.

The outline of the rest of this chapter is as follows: In Sect. 2.2 we briefly summarize the numerical methods we have used. We then discuss the phase diagram of the J-Q-h model in Sect. 2.3. In Sects. 2.4 and 2.5 we discuss metamagnetism in the J-Q-h and J_1-J_2 chains, respectively. Section 2.6 contains our results for zero-factor scaling of the saturation transition in the J-Q-h chain. In Sect. 2.7 we summarize and discuss our main results.

2.2 Methods

A more detailed description of the methods used here can be found in Chap. 5.

The primary numerical tools employed in this work are Lanczos exact diagonalization and the SSE QMC method [12] with directed loop updates [13]. Symmetries are implemented in the Lanczos calculations as described in [3]. SSE works by exactly mapping a d-dimensional quantum problem onto a $(d + 1)$-dimensional classical problem through Taylor expansion of $e^{-\beta H}$. This extra dimension is related to imaginary time in a manner similar to the path integrals in world-line QMC, but in the Monte Carlo sampling the operational emphasis is not on the paths but on the operators determining the fluctuations of the paths. We incorporate the magnetic field in the diagonal part of the two-spin (J) operators. Diagonal updates insert and remove two- and four-spin diagonal operators, while the directed loop updates change the operators from diagonal to off-diagonal and vice versa [3]. When a two-spin operator is encountered in the loop-building process, we choose the exit leg using the "no-bounce" solution of the directed loop equations for the Heisenberg model in an external field found in [13]. When encountering a four-spin Q-type operator, where the field contribution is not present, the exit leg is chosen using a deterministic "switch and reverse" strategy, essentially identical to the SSE scheme for the standard isotropic Heisenberg model [3].

When using SSE alone, we found that simulations sometimes became stuck at metastable magnetization values for long periods of time. This made it hard for simulations to reach equilibrium and difficult to compute accurate estimates of statistical errors. This problem can be easily seen in our preliminary results presented in Figs. 2 and 3 of [33], where the large fluctuations in the magnetization are due to this "sticking" problem. To remedy this, in the present work we implemented a variation of the replica exchange method [14] for QMC known as quantum replica exchange [15], implemented using the Message Passing Interface (MPI) parallel computing library.

In the traditional replica exchange method [14] (also known as parallel tempering), many simulations are run in parallel on a mesh of temperatures. In addition to standard Monte Carlo updates, replicas are allowed to swap temperatures with each other with some probability that preserves detailed balance in the extended multicanonical ensemble. This allows a replica in a metastable state to escape by wandering to a higher temperature. In the SSE simulations with replica exchange [15], we run many ($10 \sim 100$) simulations in parallel. Instead of using different temperatures as in standard parallel tempering, we use a mesh of magnetic fields. After each Monte Carlo sweep, we allow replicas to exchange magnetic fields with one another in a manner that preserves detailed balance within the ensemble of SSE configurations.

For relatively little communications overhead, we find that replica exchange can dramatically reduce equilibration and autocorrelation times, thus allowing simulations of much larger systems at much lower temperatures. In practice, adding additional replicas slows down the simulation because the time required to complete a Monte Carlo sweep varies and all the replicas have to wait for the slowest replica to finish before continuing. This slowdown can be somewhat alleviated by running more than one replica on each core.

2.3 Phase Diagram

The J-Q model has so far been of theoretical interest mainly as a tool for large-scale studies of VBS phases and AFM–VBS transitions. In a VBS (dimerized state), spins pair up to form a crystal of localized singlets, thus breaking translational symmetry but preserving spin-rotation symmetry as illustrated in Fig. 2.1a, b. The elementary quasiparticle excitations of a VBS are gapped triplet waves (triplons) formed by exciting a singlet pair to a triplet, as seen in Fig. 2.1c. Triplons sometimes deconfine into pairs of spinons: fractionalized spin-$1/2$ excitations that correspond to VBS

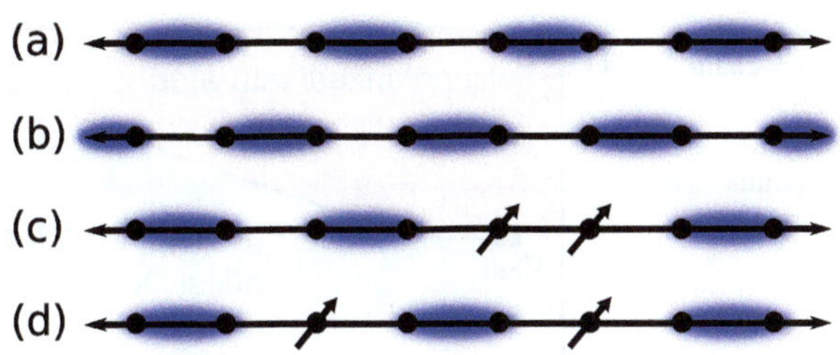

Fig. 2.1 Examples of VBS configurations of $S = 1/2$ spins in one dimension. Each blue ellipse represents a singlet pair: $(|\uparrow\downarrow\rangle - |\downarrow\uparrow\rangle)/\sqrt{2}$. In a real VBS there are also fluctuations in the singlet patterns (except in special cases) but the density of singlets on the bonds is still modulated with periodicity two lattice spacings. (**a**), (**b**) Show the two degenerate VBS ground states, (**c**) illustrates a triplet excitation in which a singlet bond is broken, and (**d**) illustrates a triplet excitation deconfined into two independently propagating spinons

domain walls as shown in Fig. 2.1d. For dimensionality $d > 1$, the spinons are confined by a string in a manner similar to quarks, the energy associated with the shifted VBS arrangement resulting from separating two spinons is directly proportional to the distance between the spinons (see [34] for a recent discussion of this analogy). In a one-dimensional VBS, the spinons are always deconfined, unless the Hamiltonian breaks translational symmetry [6, 35]. The frustrated Hamiltonians that were traditionally used to study VBS physics, e.g., the J_1-J_2 chain [29, 30, 35, 36], suffer from the sign problem, which prevents large-scale numerical simulations using QMC methods; the J-Q model is sign-problem free.

Our main aim here is to study the magnetization process of the J-Q-h chain from $h = 0$ all the way to the fully polarized state where the concept of spinons in a dimer background breaks down. To understand the basic physics in this regime, it is more appropriate to consider flipped spins ("magnons") relative to the vacuum of a fully magnetized state. For completeness, in this section we also comment on the $T = 0$ phases of the system in the full q-h plane.

Figure 2.2 shows a schematic phase diagram assembled from the literature and our own calculations. The parameter regions corresponding to the horizontal and vertical axes are well understood from past studies; the off-axes area has not been previously studied and is therefore the primary focus. The h axis is the standard Heisenberg chain in a magnetic field, where the transition into the fully polarized state is continuous. The q axis corresponds to the previously studied zero-field J-Q model [6], where for $q < q_c$ there is a Heisenberg-type critical AFM state with spin–spin correlations decaying with distance r as $1/r$ (up to a multiplicative logarithm) [37]. At $q = q_c \approx 0.8483$ the chain undergoes a dimerization transition into a VBS ground state [6]. This transition is similar to the Kosterlitz–Thouless transition and identical to the quasi-AFM to VBS transition in the J_1-J_2 chain [5, 6, 35].

In the full phase diagram for $q > 0$ (which we focus on here because $q < 0$ leads to QMC sign problems), there are three phases: a fully polarized phase, a VBS, and a partially polarized critical XY phase. If we start from a VBS state ($h = 0, q > q_c$)

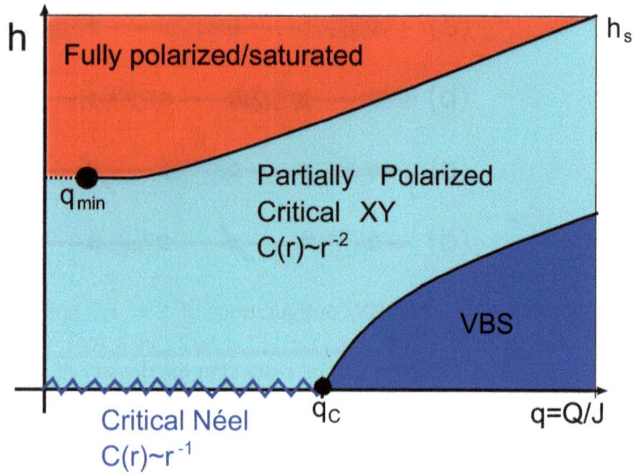

Fig. 2.2 Schematic phase diagram of the J-Q-h chain defined in Eqs. (2.2) and (2.3). The different phases and special points indicated are described in the text

and add a magnetic field, the field will "pull down" the triplet excitations with magnetization $m_z > 0$ and at some $h_c(q)$ a magnetized state becomes the ground state. These triplets originating from "broken singlets" will deconfine into spinons [6, 38], as illustrated in Fig. 2.1c, d. Each spinon constitutes a domain wall between VBS-ordered domains (as discussed in detail in [6]), and we therefore expect any finite density of spinons to destroy the VBS order. The phase boundary extending from q_c should therefore follow the gap to excite a single triplet out of the VBS. We expect the destruction of the VBS to yield a partially polarized state with critical XY correlations, as in the standard AFM Heisenberg chain in an external field. We do not focus on this part of the phase diagram here, and will not discuss the nature of the VBS–XY transition or the exact form of this phase boundary.

We focus mainly on the line $h_s(q)$ separating the XY and saturated phases in Fig. 2.2, and will provide quantitative results in the following sections. The magnetization curve is continuous along the dotted portion of h_s; here, the saturation transition is governed by a remarkably simple zero-scale-factor universality [32]. The solid portion denotes the presence of a magnetization jump: a first-order quantum phase transition known as the metamagnetic transition. The point $q_{\min}$ marks the lower metamagnetic bound, a tricritical point where the magnetization jump is infinitesimal.

2.4 Metamagnetism in the J-Q Chain

The introduction of the four-spin Q term has a dramatic effect on the magnetization process. In Fig. 2.3, we plot the magnetization density, $m(h)$, normalized to be unity in the fully polarized state,

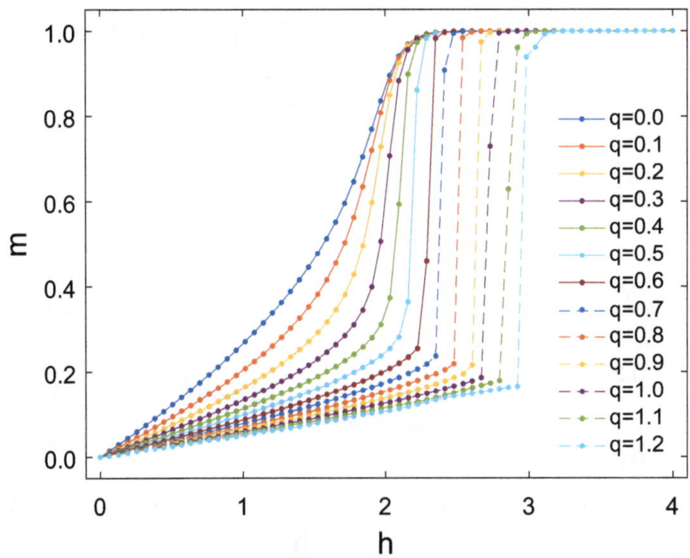

Fig. 2.3 Magnetization density of the J-Q-h chain as a function of the external field for a set of coupling ratios $0 \leq q \leq 1.2$ (from Heisenberg limit to beyond the VBS transition). The system size is $L = 96$ and the inverse temperature is $\beta = 12$ in all cases. Error bars are smaller than the markers

$$m \equiv \frac{2}{L}\sum_{i=1}^{L}\langle S_i^z\rangle, \tag{2.4}$$

for periodic J-Q-h chains with $0 \leq q \leq 1.2$, $L = 96$, and inverse temperature $\beta = 12$ (where the finite-temperature effects are already small on the scale used in the figure). We begin in the Heisenberg limit ($q = 0$) and increase q. For small q, the saturation field is unchanged, but the shape of the magnetization curve changes significantly, becoming steeper near saturation. As q increases, the magnetization seems to develop a jump to saturation and the size of this jump grows with increasing q. It is especially interesting that this jump appears for $q < q_c$, a regime where the $h = 0$ chain is in the critical AFM state and not yet in the VBS state. This magnetization jump is an example of a metamagnetic transition [16, 17] and shows many hallmarks of a first-order phase transition, including hysteresis in the QMC simulations (as documented in our earlier, preliminary paper [33]).

In Fig. 2.4 we plot the magnetization density at $q = 1.2$ for chains of sizes ranging from $L = 8$ to 256 and inverse temperature $\beta = L/4$. In this regime, we observe two distinct phases: a paramagnetic regime and a fully polarized state separated by a sharp jump. The magnetization curves exhibit near perfect agreement for all sizes studied, limited only by the discretized values of m for each size (visible in greater detail in the inset). Because of the way in which the temperature is scaled, for the smallest sizes the steps are thermally smeared out but become visible for

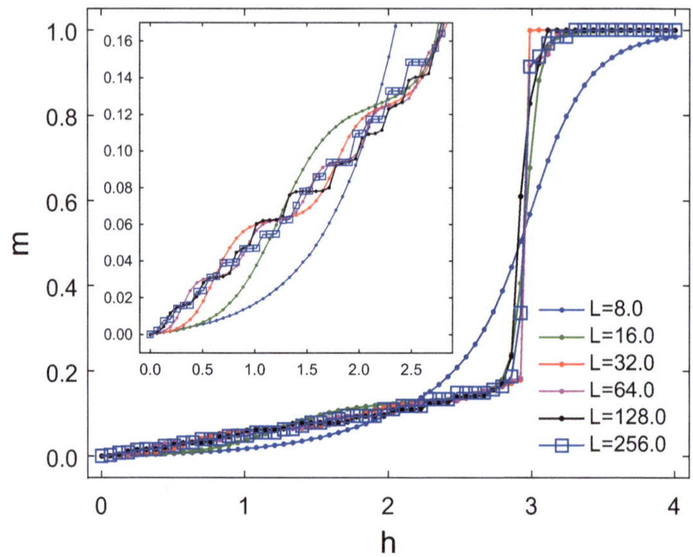

Fig. 2.4 Magnetization density of the J-Q-h chain at $q = 1.2$ as a function of the external field h, with the inverse temperature scaled with size as $\beta = L/4$. The system sizes are between $L = 8$ and 256 as indicated. The inset shows a zoomed-in view of the paramagnetic regime. The error bars are smaller than the markers in main figure and have been omitted for clarity also in the inset (where they are sometimes slightly larger than the markers)

the longer chains. Figure 2.4, as in Fig. 2.3, shows no signs of any magnetization plateaus apart from the fully polarized one. There is also no sign of the VBS gap (to the first triplet excitation), which should manifest itself as an $m = 0$ plateau for $q > q_c$, reflecting the finite field needed to close the gap. While there *is* a gap in the VBS, at these sizes and temperatures the VBS gap is too small to produce a noticeable effect. We have computed finite-size gaps using Lanczos calculations but they are difficult to extrapolate to infinite size, and we can only extract an upper bound; the triplet gap at $q = 1.2$ should be less than 0.02.[1]

It was difficult to extract precise results for the saturation field h_s or m_c (the magnetization at which the jump occurs) due to the tendency of simulations to get stuck in metastable states near the transition [33] (itself a characteristic of a first-order transition). Although the use of replica exchange has dramatically reduced this problem, it is still apparent for large chains and at lower temperatures. To extract h_s precisely, we therefore used Lanczos exact diagonalization. The external field commutes with the Hamiltonian, so we can diagonalize the zero-field J-Q model and add the contribution from the field afterwards. Figure 2.5 shows the critical magnetic field for $L = 30$ (we have also studied smaller systems in this way). For $q \leq q_{\text{min}}$, the saturation field is exactly $h_s = 2J$. In this regime, h_s is determined

[1]S. Capponi, Private communication (2017).

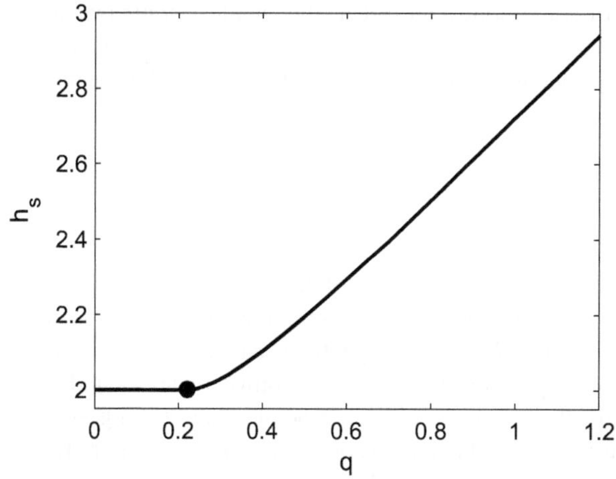

Fig. 2.5 Saturation field versus the coupling ratio for the $L = 30$ periodic J-Q-h chain calculated using the Lanczos method. The dot indicates $q_{\min}$

by a level crossing between the $m = S$ and $S - 1$ states which is independent of both q and L; see also Eqs. (A.2) and (A.10a) and (A.10b). For $q > q_{\min}$, we find a positive relationship between h_s and q, consistent with our QMC results in Fig. 2.3; here we should expect some finite-size effects, but they do not alter the qualitative character of the line $h_s(q)$.

2.4.1 Origin of the Magnetization Jump

Although the excitations of the zero-field J-Q chain are classified as spinons, near the saturation transition the density of domain walls is too high for this picture to be relevant, and the excitations are better characterized as magnons: bosonic spin-1 excitations corresponding to spin flips on a background of uniformly polarized spins. We will now show that the magnetization jump in the J-Q-h chain (and later, the J_1-J_2 chain with anisotropy in Sect. 2.5) is caused by the onset of an effective attractive interaction between these magnons.

Using an analytical approach and diagonalization of short chains, we will now derive $q_{\min}$, the minimum value of q required to produce a jump (see Fig. 2.2). This argument is described in more detail in Sect. A.1. We begin with the fact that the jump is always to the saturated state and assume that the size of the jump $\Delta m_z/L \to 0$ at $q_{\min}$ as $L \to \infty$. In an infinite system, the smallest possible jump is infinitesimal; in this case the "jump" corresponds only to a higher-order singularity (a divergence of the magnetic susceptibility). In a finite-size system, the magnetization advances by steps of $\Delta m_z \geq 1$. In a trivial paramagnet, the magnetization advances by the smallest possible increment: $\Delta m_z = 1$; this effect can be seen for $L = 256$ in the inset of Fig. 2.4. Larger magnetization steps indicate the presence of some nontrivial effect; the smallest *nontrivial* jump is $\Delta m_z = 2$, i.e., a direct level crossing between $m_z = S - 2$ and S. In Sect. A.1, we discuss the details of a two-magnon approach to solving this problem using the condition for the level crossing:

$$\bar{E}_2 \leq 2\bar{E}_1, \tag{2.5}$$

where $\bar{E}_n$ is the zero-field n-magnon ground state energy as defined in Eq. (A.10).

Equation (2.5) essentially requires that the interaction between the magnons be attractive, since the energy of two interacting magnons is *lower* than twice the energy of a single magnon. Metamagnetism can be brought on by the appearance of bound states of magnons if there is an instability toward bound states of ever more magnons. Thus, the existence of such a bound state is suggestive of, but does not guarantee, the existence of a macroscopic magnetization jump. If the bound pairs of magnons are not attracted to other bound pairs of magnons, then the magnetization merely advances by steps of $\Delta m_z = 2$ without any macroscopic jump. This effect has been documented previously [23, 39]: in a liquid of bound states of two or more magnons, the magnetization undergoes microscopic jumps where Δm_z is an integer equal to the number of bound magnons, with in principle, an infinite number of such phases existing, but never a macroscopic jump.

Thanks to the QMC data, there can be no doubt of the existence of a macroscopic magnetization jump in the J-Q-h chain for $q > q_{\min}$, but it would be difficult to extract an accurate value for $q_{\min}$ from these data alone. Instead, we will determine a precise value of $q_{\min}$ using the condition in Eq. (2.5). To do this, we first note that the effect of the Q term on the two-magnon subspace is a short-range *attractive* interaction, albeit an unusual one including correlated hopping (see Sect. A.1 for a detailed analysis). From Eq. (A.2) we know that $\bar{E}_1 = -2J$ and we can then find a condition on $\bar{E}_2$ for a bound state to form as a result of this attraction:

$$\bar{E}_2 \leq -4J. \tag{2.6}$$

With this in hand, we may interpret the magnetization jumps seen in the QMC data for $q > q_{\min}$ as follows: At higher magnetization densities, this short-range attractive force dominates, causing the gas of magnetic excitations to suddenly condense, producing a magnetization jump. Indeed, when the magnetization was fixed at a nonequilibrium value in the QMC calculations (for example, $m = 1/2, q = 1.2$), we observed phase separation: the chain would separate into a region with magnetization density m_c and another region that was fully polarized. Therefore, we may identify $q_{\min}$ with the threshold value of q at which Eq. (2.6) is first satisfied.

In Fig. 2.6 we plot $\bar{E}_2(J = 1, Q = q)$; we can determine $q_{\min}$ by finding the smallest value of q that satisfies Eq. (2.6). In this way, we obtain $q_{\min} = 0.\bar{2} = \frac{2}{9}$ to machine precision for all $L > 6$. For $q < q_{\min}$, finite-size effects result in an *overestimate* of $\bar{E}_2(L \to \infty)$, and for $q > q_{\min}$, they result in an *underestimate*.

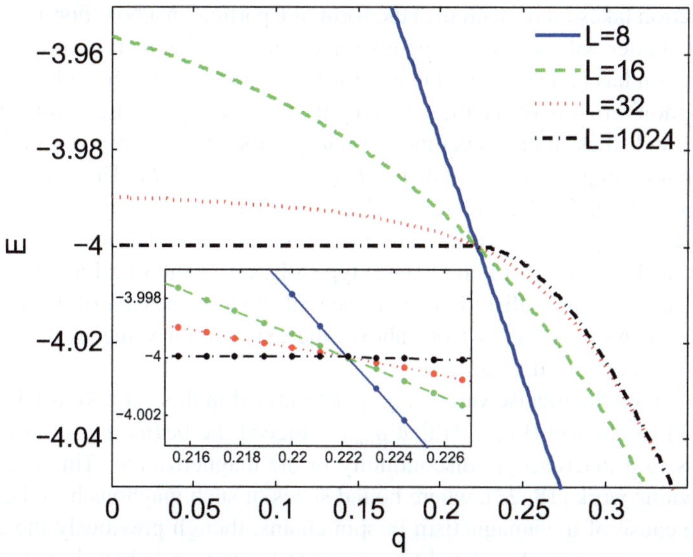

Fig. 2.6 The lowest-energy eigenvalue $\bar{E}_2(J = 1, Q = q, L)$ in the two-magnon sector ($m_z = S - 2$) in the J-Q-h chain for system sizes $L = 8, 16, 32, 1024$

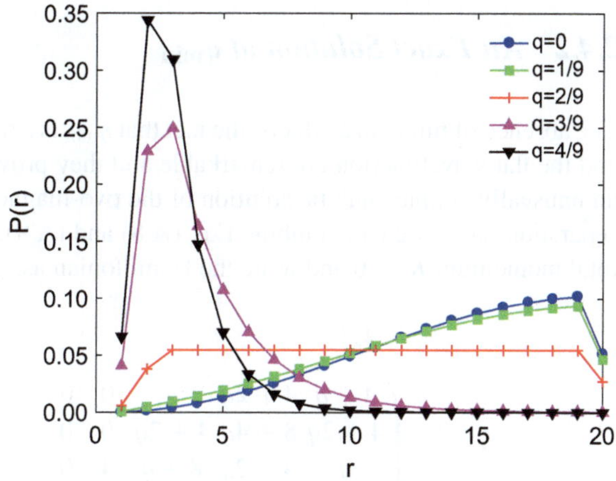

Fig. 2.7 The probability $P(r) = \langle\psi_0(r)|\psi_0(r)\rangle$ of the particles being separated by distance r in the lowest state in the two-magnon sector ($m_z = S - 2$) of the J-Q-h chain

At exactly $q = q_{\min}$, these effects cancel and $\bar{E}_2$ becomes independent of L (for $L > 6$). Note that $q_{\min} < q_c$ (the VBS critical point); we should not expect q_c and $q_{\min}$ to match since the magnetization jump occurs not from the VBS but from the critical XY state and they are arise from completely different mechanisms.

In Fig. 2.7, we plot the probability density $|\psi_0(r)|^2$ for $L = 40$ chains at several values of q (r is the magnon separation in the separated center-of-mass and relative-coordinate basis as defined in detail in Sect. A.1). For $q < q_{\min}$, the magnons scatter off one another with a finite-range effective repulsive interaction, and the relative wavefunction takes on (essentially) the form of a particle in a box. For $q > q_{\min}$, the magnons scatter with a finite-range effective *attractive* interaction, in this case the wavefunction has an exponential decay for $r \geq 3$, indicating a bound state. At $q = q_{\min}$, magnons cross between these two regimes, scattering off one another acquiring no phase and, thus, their wavefunction and ground state energy resemble that of two noninteracting magnons, with $\bar{E}_2(J, Q_{\min}) = 2 \cdot (-2J)$. The wavefunction is exactly constant in the bulk ($3 < r < L/2 - 1$). This completely flat wavefunction in the bulk at $q_{\min}$ (which we will discuss analytically further later) is *not* a generic behavior at the onset of a bound state; typically, one would find an exponentially decaying short-distance disturbance (as we will show in one case of the J_1-J_2 chain in Sect. 2.5). As $q \to q_{\min}$ from above, the expectation value of the separation between the magnons diverges.

Finally, with the precise value of $q_{\min}$ determined in this way, we use large-scale QMC data to confirm (Fig. 2.3) that $q_{\min}$ is indeed the beginning of an instability that leads to a macroscopic discontinuity in the magnetization. This is consistent with previous work [18, 23], where bound states of such magnons have been found to be the cause of metamagnetism in spin chains, though previously the attractive interactions were directly related to geometric frustration (which is not present in the J-Q chain; the Q term competes in a different way against AFM order).

2.4.2 An Exact Solution at $q_{\min}$

The absence of finite-size effects, the fact that $q_{\min}$ is a ratio of small whole numbers, and the flat wavefunction are remarkable and they provide a hint that there may be an unusually simple analytic solution of the two-magnon system at $q_{\min}$. Using the separation basis, we can combine Eqs. (A.8) and (A.9), set $J = 1$, $Q = q$, and the total momentum $K = 0$ and write the Hamiltonian as:

$$-4H = \tag{2.7}$$

$$\begin{pmatrix} 4+q & 4+2q & q & 0 & 0 & \cdots & 0 \\ 4+2q & 8+4q & 4+2q & 0 & 0 & \cdots & 0 \\ q & 4+2q & 8+q & 4 & 0 & \cdots & 0 \\ 0 & 0 & 4 & 8 & 4 & 0 & \cdots \\ \vdots & \vdots & \ddots & \ddots & \ddots & \ddots & \ddots \\ \vdots & \vdots & 0 & 4 & 8 & 4 & 0 \\ \vdots & \vdots & \ddots & 0 & 4 & 8 & 4\sqrt{2} \\ 0 & 0 & \cdots & \cdots & 0 & 4\sqrt{2} & 8 \end{pmatrix}.$$

Using the simple-looking numerical result for the wavefunction $\psi(r, q = q_{\min})$ in Fig. 2.7 as inspiration for finding the ground state, we will now assume (and later confirm) that it has the following form:

$$|\psi\rangle \propto +a\,|1\rangle + b\,|2\rangle + c\,|3\rangle + \sum_{r=4}^{L/2-1} |r\rangle + d\,|L/2\rangle\,. \tag{2.8}$$

The wavefunction is constant in the bulk, but at the edges of the r subspace the state has weights a, b, c, d that can be easily determined. Acting on $|\psi\rangle$ with H in Eq. (2.7) produces a set of five equations which can be solved for a, b, c, d, q_{min} and the eigenvalue λ with the following results:

$$a = \frac{1}{3},\quad b = \frac{5}{6},\quad c = 1,\quad d = \frac{1}{\sqrt{2}}, \tag{2.9a}$$

$$\lambda = -4J, \tag{2.9b}$$

$$q_{\text{min}} = \frac{2}{9}. \tag{2.9c}$$

When this solution is plugged back into Eq. (2.8), we indeed find an exact match for the numerical results for $q = q_{\text{min}}$ plotted in Fig. 2.7.

2.4.3 Excluded Mechanisms for Metamagnetism

We will now discuss some other processes known to cause magnetization jumps, such as localization [40–42], magnetization plateaus [43], and multipolar phases [44] and then show that none explain the behavior of the J-Q-h chain. Although metamagnetism *can* be caused by localization [40–42], this cannot be the cause in this case because the J-Q-h chain has no intrinsic disorder and we see no other signs of localization. Metamagnetism has also been observed in a study of the frustrated FM Heisenberg chain [19, 24, 44], which has a sequence of multipolar phases. If such phases existed near q_{min}, we would observe a "cascade" of jumps. First, the smallest possible jump of $\Delta m_z = 2$ would appear, but then for slightly larger values of $q > q_{\text{min}}$, there should be a series of system-size-independent jumps, $\Delta m_z = 3, 4, 5, \ldots$ until, eventually, a macroscopic jump in the thermodynamic limit. Based on exact diagonalization of chains up to $L = 28$, we see no evidence of such size-independent jumps in the J-Q-h chain nor do we see any evidence of such an effect in our QMC data.

A jump in the magnetization can also be connected to a magnetization plateau [43]. There is no sign of a magnetization plateau in Figs. 2.3 or 2.4, but to conclusively rule this out, we can also examine spin correlation functions. A magnetization plateau indicates the presence of a gap between different spin states and is allowed (by an extension of the Lieb–Schultz–Mattis theorem) only when the magnetization per unit cell, m, obeys the constraint that $(S - m)$ is an integer

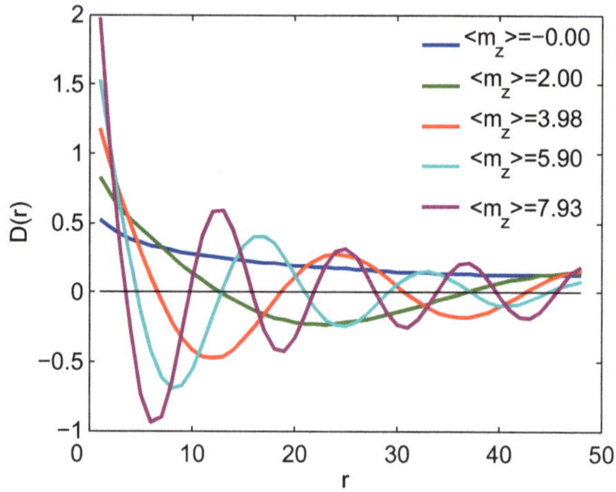

Fig. 2.8 Alternating dimer–dimer correlation function, defined in Eq. (2.10), for several values of the magnetization in chains of length $L = 96$ at $\beta = 12$, $q = 1.2$

[45]. For an $S = 1/2$ chain, this can only occur by breaking translational symmetry. We examined the alternating part of the dimer–dimer correlation function, $D(r)$, for signs of translational symmetry breaking. This correlation function is defined as

$$D(r) = (-1)^r \left[B(r) - B(r+1)\right], \tag{2.10}$$

where $B(r) = \langle P_{i,i+1} P_{i+r,i+1+r} \rangle$ measures the correlations between bond singlet densities. In the VBS-ordered phase, $D(r)$ has the form $D(r) \propto (e^{-r/\xi} + D_0)$, where D_0 is the VBS order parameter. In Fig. 2.8, we plot $D(r)$ for several different values of the magnetization. For $m_z > 0$, $D(r)$ develops long-wavelength oscillations with a wavelength proportional to the inverse magnetization density $\lambda \propto 1/m$ (a similar effect was predicted in 1D quantum fluids by Haldane [46]), but we find no evidence of broken symmetry. The S^z spin correlations develop a similar pattern of long-wavelength oscillations, and also show no signs of a symmetry-broken state. As a final test, we looked at chains with open boundaries and found no signs of symmetry-broken states in that case either.

2.5 Metamagnetism in the J_1-J_2 Chain

In the J-Q-h chain, the Q term favors AFM ordering at the classical level (where the singlet projection aspect is not manifested), but nonetheless it produces a short-range *attractive* interaction for low densities of magnons (against a saturated background). Other Hamiltonians with these features may exist, and since they also lack frustration, they are likely to be understudied. Using the recipe from the J-Q-h chain: (AFM first-neighbor interaction) + (short-range attractive magnon–magnon interaction), a natural challenge is then to create a minimal unfrustrated quantum spin model which also exhibits this effect using only two-spin interactions. We can construct a minimal model by adding an anisotropic *ferro*magnetic (FM) next-nearest neighbor term to the AFM Heisenberg chain. We will now show that a bound state of magnons occurs in the J_1-J_2 model, but only with spin anisotropy in the J_2 term, i.e., with the Hamiltonian

$$\begin{aligned} H_{J_1 J_2} = &- J_1 \sum_i P_{i,i+1} \\ &- J_2 \sum_i \left[\frac{1}{4} - S_i^z S_{i+2}^z - \frac{\Delta}{2} \left(S_i^+ S_{i+2}^- + H.c. \right) \right]. \end{aligned} \tag{2.11}$$

Here, we have defined Δ in such a way as to guarantee that the $S^z S^z$ interactions of the second-neighbor term are FM for all $J_2 < 0$.

When $\Delta = 1$, $J_2 > 0$ (AFM), Eq. (2.11) becomes the simplest example of a frustrated spin model; this case has been well studied [18–20, 23–25, 29–31, 47–49]. Several papers have presented evidence of metamagnetism in the J_1-J_2 chain in this regime for both the isotropic [19, 22–25] and anisotropic [18, 20, 21, 25] cases. Naively, a FM second-neighbor term is trivial since it does not produce frustration; with an AFM first-neighbor coupling it would serve to strengthen the AFM order. Probably for this reason, the FM J_2 case has been almost completely overlooked in the literature. Only a few papers [48–50] have considered this case and none of them investigated the possibility of metamagnetism. Metamagnetism *has* been reported in the 2D and 3D AFM Ising model with a FM second-neighbor term [26], and a physically equivalent square-lattice-gas model [27].

As with the J-Q-h chain, we will identify the onset of a bound state of two magnons on a fully polarized FM background. As we discussed in Sect. 2.4.2, such a bound state is a possible signature of metamagnetism, but not a guarantee of it (although in any case the onset of a bound state is an important aspect of other possible transitions). We define the criterion for the bound state as

$$\bar{E}_2(j, \Delta) \leq 2\bar{E}_1(j, \Delta), \tag{2.12}$$

where $J_1 = 1$ (AFM), $j \equiv -J_2/J_1$ ($j > 0$ corresponding to FM J_2). The magnon binding energy is therefore

$$\Xi(j, \Delta) \equiv 2\bar{E}_1 - \bar{E}_2, \tag{2.13}$$

such that $\Xi > 0$ indicates the presence of a bound state.

The exact one-magnon energy, $\bar{E}_1$, is derived in Sect. A.2 and displayed in Eq. (A.20). The two magnon energy, $\bar{E}_2$, can be determined numerically using the separation basis Hamiltonian constructed from H_{J_1} and H_{J_2} [Eqs. (A.21) and (A.22)]. We will limit ourselves to the unstudied case of FM J_2 ($j > 0$) and, for simplicity, we will consider only three values of Δ: $\Delta = 1$ (the isotropic case);

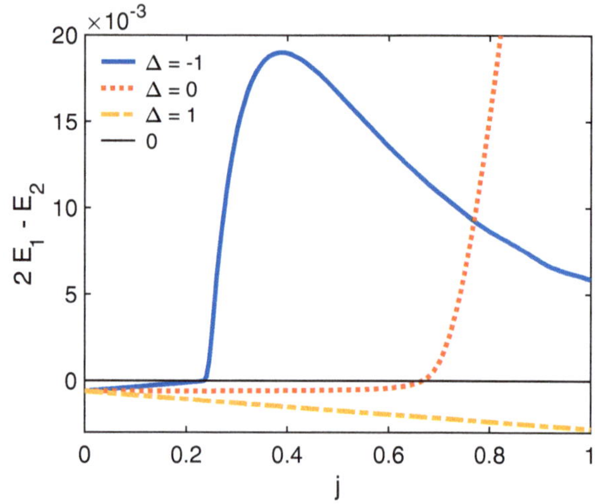

Fig. 2.9 The binding energy defined in Eq. (2.13) for a J_1-J_2 chain with $j \equiv -J_2/J_1$ and anisotropy parameters $\Delta = 0, \pm 1$. Here a relatively small system ($L = 128$) is used, to make it easier to see the crossings. When $\Xi(j, \Delta) > 0$, there is a bound state of two magnons

$\Delta = 0$ (the Ising case); and $\Delta = -1$ (where the Ising interaction is FM and the XY interactions are AFM).

In Fig. 2.9, we plot $\Xi(j, \Delta)$ versus j for chains of length $L = 128$. For large L, the level crossing occurs at a very shallow angle and the lines in Fig. 2.9 tend to overlap; we therefore use a small system size here to make the crossing more clear. In the isotropic case, $\Delta = 1$, $\Xi(j, 0) < 0$ for all j and there is no bound state. In the Ising case, $\Delta = 0$, there is a level crossing at $j_{\min} = \frac{2}{3}$ (verified to machine precision for chains up to $L = 4096$), and for $\Delta = -1$ the bound state occurs above $j_{\min} = 0.236067977499$ (to machine precision for $L \geq 32$).

For $\Delta = 0$, the wavefunction takes on a flat form at $j_{\min} = \frac{2}{3}$. Using the same approach we used for $q_{\min}$ in Sect. 2.4.2

$$|\psi\rangle \propto -\frac{1}{3}|1\rangle + \sum_{r=2}^{L/2-1} (-1)^r |r\rangle + \frac{1}{\sqrt{2}} |L/2\rangle . \tag{2.14}$$

Except for the alternating sign, this is almost identical to the flat wavefunction for the J-Q-h chain at $q_{\min}$ and finite-size effects are similarly absent at this point. For $\Delta = -1$, the form for the ground state at $j_{\min}$ is nearly flat with an exponential tail,

$$|\psi\rangle \propto \sum_{r=1}^{L/2-1} (-1)^r (1 - ae^{-r/b}) |r\rangle + \frac{(-1)^{L/2}}{\sqrt{2}} |L/2\rangle , \tag{2.15}$$

where $a = 1.447$ and $b = 2.078$, based on a fit to the numerical wavefunction (solving directly involves a transcendental equation that we have not studied further). In this case, finite-size effects are present, but vanish exponentially in L. The existence of this two-magnon bound state may be a precursor to a macroscopic magnetization jump, but there is no guarantee that it produces the required instability to multi-magnon bound states. Confirming the existence of this transition with large-scale calculations would be an interesting topic for a future study, although the regime $\Delta < 0$ is inaccessible to QMC due to the sign problem.

2.6 Zero-Scale-Factor Universality

The critical behavior that has become known as zero-scale-factor universality occurs when response functions are universal functions of bare coupling constants with no nonuniversal factors [32]. Zero-scale-factor universality is expected to apply in one-dimensional systems whenever there is a continuous quantum phase transition that corresponds to the smooth onset of a nonzero ground state expectation value for a conserved density variable. In spin chains, the most well-studied realization is the field-tuned transition from the Haldane-gapped singlet state of integer spin chains to a state in which one polarization of triplet magnons ($S = 1$ quasiparticle excitations above the singlet state) condenses to give a nonzero magnetization density.

The saturation transition in the J-Q-h chain provides a different realization: the magnons are now single spin-flip excitations above the saturated (i.e., fully polarized) ground state (the same magnons as in Sect. 2.4.2), and the transition in question is the transition from the saturated state to the partially polarized critical state. When this transition is continuous the density of magnons turns on continuously. Moreover, the density of these magnons is conserved by virtue of the U(1) symmetry of spin rotations about the z axis. Therefore, the magnetization density, Eq. (2.4), in the vicinity of the saturation transition, is expected to obey the following form [from Eq. (1.23) of [32]]:

$$\langle m \rangle = g\mu_B \left(\frac{2M}{\hbar^2 \beta}\right)^{1/2} \mathcal{M}(\mu\beta), \tag{2.16}$$

where M is the magnon mass and $\mu = (h_s - h)$.

The single magnon dispersion (A.2) obeys the low-energy quadratic form $\epsilon(k) \propto k^2/(2M)$, with $M = 1$ (in our units where $J = 1$) independently of Q. The Q term gives rise to an additional contribution to the hopping if two magnons are within three lattice spacings of each other. Considering the low magnon density and repulsive magnon–magnon interactions, we only expect a negligible renormalization of M due to this correlated hopping term. We define $\langle m \rangle = g\mu_B \langle n \rangle$, where n is the density of flipped spins and $\mu = (h_s - h)$. In this way, the field above the saturation value represents the "gap" for these magnetic excitations and a negative μ corresponds to $h > h_s$. We insert these definitions into Eq. (2.16):

$$\langle n \rangle \left(\frac{\hbar^2 \beta}{2M} \right)^{1/2} = \mathcal{M}[\beta(h_s - h)] \tag{2.17}$$

To simplify further, we set $\hbar = 1$ and define the rescaled field $t \equiv \beta(h_s - h)$:

$$n_s(q, t) \equiv \langle n \rangle \sqrt{\frac{\beta}{2M}} = \mathcal{M}(t) \tag{2.18}$$

We will henceforth call n_s the rescaled magnon density. The one-dimensional case is unique here, in that there is a known analytic form [32] for the universal scaling function $\mathcal{M}(t)$:

$$\mathcal{M}(t) = \frac{1}{\pi} \int_0^\infty dy \frac{1}{e^{y^2 - t} + 1} = -\frac{1}{2\sqrt{\pi}} \mathrm{Li}_{1/2}(-e^t) \tag{2.19}$$

In the limit $|t| \to \infty$, the polylogarithm simplifies and the universal function becomes

$$\mathcal{M}(t) = \begin{cases} \frac{\sqrt{t}}{\pi} & t \to \infty, \\ \frac{e^t}{2\sqrt{\pi}} & t \to -\infty, \end{cases} \tag{2.20}$$

but we will use the full form without approximations.

The critical behavior of the magnetization near the saturation field was recently studied using the finite-temperature Bethe ansatz in the case of the standard $S = 1/2$ Heisenberg chain [51], and detailed comparisons were also made with experimental results for AFM chain [52, 53] and ladder [54] systems. In order to explicitly test the validity of the zero-scale-factor universality, we here analyze our data in a different manner from [51].

In Fig. 2.10, we plot the rescaled density, n_s, as a function of the rescaled field, t, for $L = 96$ J-Q-h chains near the saturation transition for $q = 0$, 0.1, 0.2 and $q = q_{\min}$. In all these cases, $h_s = 2J$ (see Fig. 2.5). The rescaled data collapse reasonably well for $q = 0$, as shown in Fig. 2.10a, although it is also clear that we have not quite reached the asymptotic large-β scaling limit (the curves for even the highest β values still exhibit some drift). We have investigated other system sizes to ensure that finite-size corrections are not important here (see also Fig. 2.11). Figure 2.10b–d, we apply the same rescaling and find that the agreement with the theory becomes progressively worse for increasing q. The curves for different temperatures still collapse rather well onto one another for $t < 0$, but the collapsed data no longer match the shape of the universal function, even if we choose M different from the bare value $M = 1$ in the single-magnon dispersion (and, as already noted, we do not expect any significant renormalization of M due to many-body effects at these low magnon densities). Additionally, the quality of the collapse

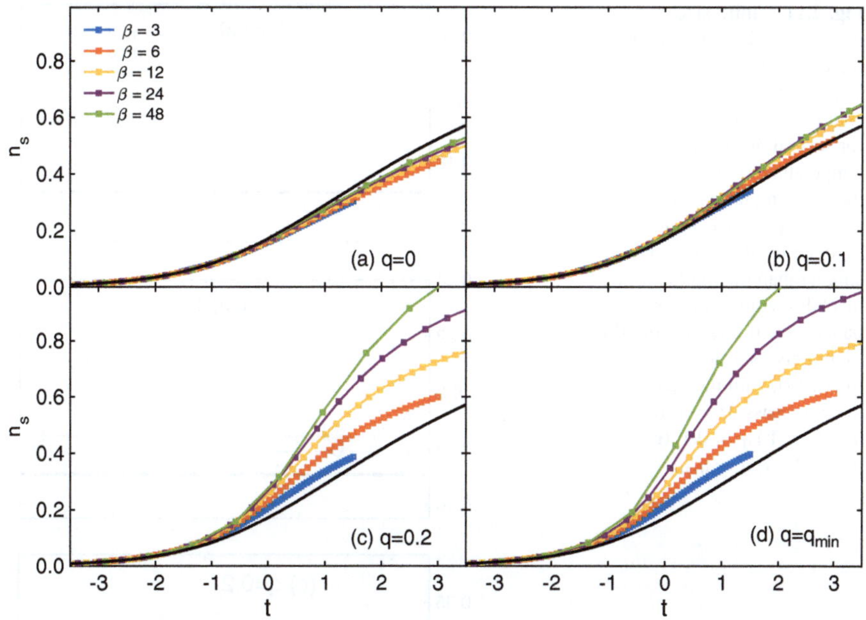

Fig. 2.10 Test of zero-factor scaling using the rescaled density, Eq. (2.18) of flipped spins near saturation for a J-Q-h chain of 96 sites for several different inverse temperatures β and values of the coupling ratio q (in different panels as indicated). The results are graphed versus the rescaled magnetic field $t \equiv \beta(h_s - h)$. The black lines are the exact predicted universal function, Eq. (2.19) with the bare magnon mass $M = 1$. (**a**) $q = 0$. (**b**) $q = 0.1$. (**c**) $q = 0.2$. (**d**) $q = q_{min}$

itself deteriorates for $t > 0$. As expected, for $q > q_{\min}$ (not shown) the zero-factor scaling fails completely: the magnons now interact attractively, and there is discontinuity in $\langle n \rangle$ which cannot be rescaled to match an analytic function.

It is not obvious from Fig. 2.10 that this scaling form works at all for $q \neq 0$. To explore this more carefully, we examine the finite-size scaling of n_s with the field set to saturation ($t = 0$) in Fig. 2.11. In this case, the exact universal function has no dependence on β, but in all panels of Fig. 2.11, there remains significant β dependence. Clearly, we have not yet reached the low temperatures (high β) where the universal form applies without significant corrections (as seen in Fig. 2.12, exceedingly low temperatures are required to observe this convergence, especially for $q > 0$). The β dependence becomes stronger for larger values of q. We also see non-monotonic β-dependence for $q = 0.1$ and 0.2, which manifests as the crossing of lines in Fig. 2.11b, c. This non-monotonic behavior explains why, in Fig. 2.10b, c, the agreement with the exact function sometimes gets *worse* for increasing β. At $q = q_{\min}$ the agreement with the exact form is far worse and n_s at $t = 0$ shows no signs of convergence. Instead, it shows a monotonic increase with β; this supports the notion that $q_{\min}$ is a tricritical point with a different scaling behavior. The cross-overs seen in the β-dependence for $0 < q < q_{\min}$ should then be due to a cross-over temperature related to the tricritical point.

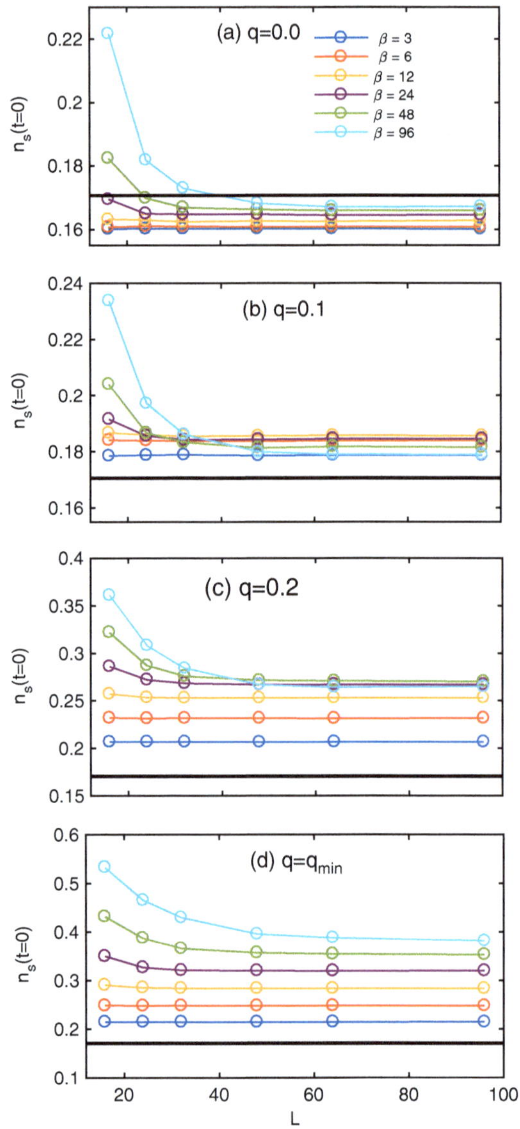

Fig. 2.11 Finite-size behavior of the zero-factor scaled magnon density, Eq. (2.18), for the J-Q-h chain at $t \equiv \beta(h_s - h) = 0$ for several different inverse temperatures β and values of the coupling ratio q (in different panels as indicated). In all cases, the error bars are smaller than the markers. The black horizontal lines in each panel show the value from the exact universal function, Eq. (2.19), with the bare magnon mass $M = 1$. (**a**) $q = 0.0$. (**b**) $q = 0.1$. (**c**) $q = 0.2$. (**b**) $q = q_{min}$

We take a closer look at the temperature dependence in Fig. 2.12, where we plot n_s at $t = 0$ versus the temperature $T = \beta^{-1}$ for a fixed size $L = 96$. Here, the cross-over behavior is clear and we know from Fig. 2.11 that finite-size effects are not important at this size. The dashed black line represents the exact value of the universal function from Eq. (2.19) evaluated at $t = 0$, $M = 1$. For $q = 0$, we can see that the results converge monotonically toward the expected value from below. With $q = 0.05$, $n_s(t = 0)$ is extremely close to the exact value, but a careful examination shows that the behavior is non-monotonic with a broad maximum before a flattening

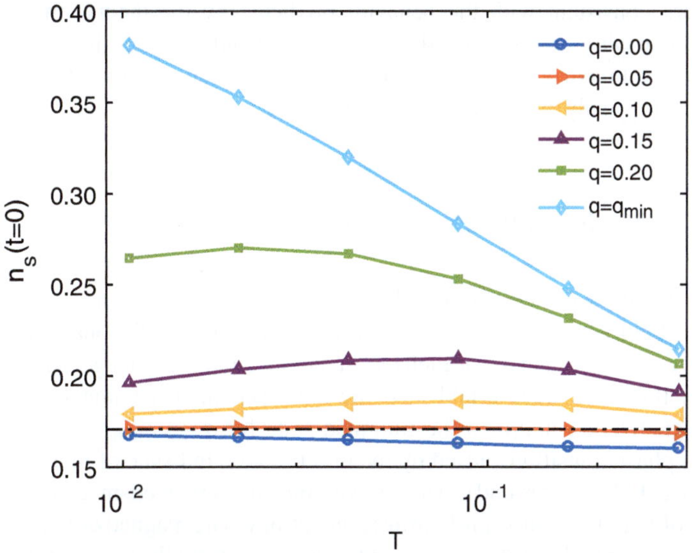

Fig. 2.12 Temperature dependence of the rescaled magnon density, Eq. (2.18), for an $L = 96$ J-Q-h chain at $h = h_s$ and several values of the coupling ratio q. Error bars are smaller than the markers. The black dashed line shows the exact asymptotic ($T \to 0$) value from the universal function, Eq. (2.19), setting the bare magnon mass $M = 1$

out at lower temperatures, consistent with asymptotic convergence to the expected value. For $q = 0.1$, the behavior of $n_s(t = 0)$ is similar and more clearly visible on the scale of Fig. 2.12. For $q = 0.15$, 0.2, there is a maximum at lower T but we cannot see the convergence to the universal value when $T \to 0$, although we expect this to take place at still lower temperatures. For $q = q_{\min}$, the behavior is essentially a logarithmic divergence, but we do not know the power of the logarithm. All these behaviors are consistent with a low-energy description with a $|\psi|^4$-type field theory, where the coefficient of the $|\psi|^4$ term vanishes at $q_{\min}$, and at this point the critical behavior is controlled not by the zero-scale-factor theory but by the marginal $|\psi|^6$ term (causing the logarithmic scaling). The cross-over temperature between the two critical behaviors, as manifested by the maximum in $n_s(t = 0)$ versus T, should gradually approach $T = 0$ as $q \to q_{\min}$ from below, as we indeed observe in Fig. 2.12.

We summarize our findings on the zero-scale-factor universality. In Fig. 2.10, we observe that this scaling works very well for $q = 0$, but the scaling appears to work poorly for $0 < q \leq q_{\min}$. By examining finite-size scaling of the rescaled magnetization in Fig. 2.11, we observe non-monotonic temperature dependence for $0 < q < q_{\min}$. Finally, in Fig. 2.12, we plot n_s as a function of T for $t = 0$, here we can see that for all $q < q_{\min}$, n_s appears to converge toward the exact value at $T \to 0$. As q approaches $q_{\min}$, the temperature required to observe convergence becomes extremely low due to the influence of the tricriticality. These results are consistent with the behavior predicted by the theory: the zero-scale-factor universality applies for all $q < q_{\min}$ and fails only at the tricritical point $q_{\min}$. Finally, this divergence occurs for $q_{\min} = \frac{2}{9}$ which confirms the results of the level-crossing analysis documented in Sect. 2.4.2.

2.7 Conclusions and Discussion

In this chapter, we have studied the J-Q chain in the presence of an external magnetic field using range of techniques including exact diagonalization, a few-magnon expansion, and a parallelized quantum replica exchange within the SSE QMC method. We have established the existence of a metamagnetic transition (i.e., magnetization jump) to the saturated state for $q \geq q_{\min} = \frac{2}{9}$, a first-order quantum phase transition caused by the onset of a bound state of magnons (flipped spins on a FM background). This proves that metamagnetism can occur in the absence of both frustration *and* intrinsic anisotropy. The magnetization jump begins with zero magnitude at $q = q_{\min}$ and increases gradually in magnitude with q. Below $q_{\min}$, magnons interact with a finite-range effectively repulsive interaction. Above $q_{\min}$, magnons interact with a finite-range effectively *attractive* interaction, despite the absence of any explicitly FM interactions. At the onset of the jump, magnons become noninteracting (for sufficiently low density) and the problem of two magnons in a polarized background can be solved analytically. The point at which two magnons bind represents the onset of an instability where an arbitrary number of magnons attract to form a macroscopic magnetization jump. Motivated by the work presented here, the existence of metamagnetism in the J-Q-h chain and our proposed mechanism for it have been confirmed by calculations using the density matrix renormalization group [55].

It may be difficult to find an experimental realization of the J-Q model itself, but interactions similar to the Q term can appear in effective models of spin-phonon chains where the phonons have been integrated out [56]. Thus, spin-phonon systems may possibly harbor metamagnetism even in the absence of longer-range frustrated Heisenberg exchange interactions. We again stress that $q_{\min}$, the threshold for metamagnetism, is significantly smaller than q_c, the threshold for dimerization; therefore, spin-phonon systems may also harbor metamagnetism even if the spin-phonon coupling is insufficiently strong to produce dimerization [57].

The saturation transition in the J-Q-h chain is rich, and we have shown that the magnetization near saturation obeys a zero-scale-factor universality [32] at $q = 0$, which becomes increasingly difficult to observe as q is increased above about ≈ 0.1. This is explained by the influence of the tricritical point at $q_{\min}$, where the low-energy effective field theory changes, leading to a different criticality and cross-over behavior. The most natural scenario is that the coefficient of $|\psi|^4$ vanishes in the $|\psi|^4$ effective field theory for the saturation transition at the threshold for formation of the two-magnon bound state, thereby allowing the $|\psi|^6$ term to control the scaling behavior of the saturation transition at this threshold. This term is marginal in spatial dimension $d = 1$ since the dynamical exponent for the transition is $z = 2$, implying the presence of logarithmic violations of scaling at $q = q_{\min}$. In our QMC data, we indeed observe logarithmic scaling of the magnetization density exactly at $q_{\min}$.

Using the same two-magnon approach from the J-Q-h chain, we have studied the AFM-FM J_1-J_2 chain with anisotropy Δ in the J_2 term [see Eq. (2.11)]. We have found that for $\Delta = 0, -1$, there is a bound state of magnons for $j > j_{\min}$ with $j_{\min} = \frac{2}{3}$, 0.236, respectively. It is likely that these bound states will cause a magnetization jump to saturation in this model, but we have not investigated this possibility using large-scale simulations. The S^z interactions in the J_2 term are in both cases FM and have the effect of reinforcing the zero-field ground state correlations. Thus, they produce no frustration in the conventional sense, but still lead to nontrivial behavior. To the best of our knowledge, no study has previously attempted to find metamagnetism in the AFM-FM J_1-J_2 chain, and this would be an excellent topic for a future study using the density matrix renormalization group method, which is well suited for frustrated one-dimensional systems. Such a study could also confirm whether the zero-scale-factor universality is obeyed by the J_1-J_2 chain near saturation and compare the breakdown as $j \to j_{\min}$ to the breakdown that occurs in the J-Q-h chain. Indeed, the AFM-FM J_1-J_2 chain may be generally understudied due to its lack of conventional frustration. The existence of a nontrivial behavior in this previously overlooked unfrustrated spin chain may mean that there are other phenomena to explore in such naively trivial Hamiltonians.

The methods developed for this work, including the parallelized replica exchange quantum Monte Carlo program, have been extended to study the 2D J-Q-h model in the presence of a magnetic field. This study, presented in Chap. 3, demonstrates that in 2D there are also magnetization jumps above a coupling ratio $q_{\min}$ and a similar mechanism of bound states of magnons as in one dimension. In two dimensions we do not expect zero-scale-factor universality close to saturation for $q < q_{\min}$, because we are then at the upper critical dimension (2+2) of this theory. Logarithmic corrections may then be expected for all $q < q_{\min}$, and the behavior at $q_{\min}$ is unclear at present.

The lower metamagnetic bound, $q_{\min}$, is less than q_c (the dimerization transition), and indeed, the physics of metamagnetism appears completely unrelated to the physics of the dimerization transition. More generally, we note the utility of J-Q-type models for studies of phenomena normally associated with frustration due to competing exchange interactions, e.g., J_1-J_2 Heisenberg models. Due to the absence of sign problems, these models can be studied with QMC simulations in any number of dimensions, while techniques for frustrated models (e.g., the density matrix renormalization group technique) are restricted to one-dimensional and relatively small two-dimensional systems. VBS physics, in particular the AFM–VBS transition, has so far been the primary goal of studies with J-Q models, and our present work now adds metamagnetism and high-field scaling to this repertoire of phenomena accessible to QMC simulations of this family of "designer Hamiltonians" [4].

References

1. A. Iaizzi, K. Damle, A.W. Sandvik, Phys. Rev. B **95**, 174436 (2017). https://doi.org/10.1103/PhysRevB.95.174436
2. A.W. Sandvik, Phys. Rev. Lett. **98**, 227202 (2007). https://doi.org/10.1103/PhysRevLett.98.227202
3. A.W. Sandvik, in *American Institute of Physics Conference Series*, ed. by A. Avella, F. Mancini, vol. 1297 (2010), pp. 135–338, http://arxiv.org/abs/1101.3281. http://dx.doi.org/10.1063/1.3518900
4. R.K. Kaul, R.G. Melko, A.W. Sandvik, Annu. Rev. Condens. Matter Phys. **4**, 179 (2013). https://doi.org/10.1146/annurev-conmatphys-030212-184215
5. S. Sanyal, A. Banerjee, K. Damle, Phys. Rev. B **84**, 235129 (2011). https://doi.org/10.1103/PhysRevB.84.235129
6. Y. Tang, A.W. Sandvik, Phys. Rev. Lett. **107**, 157201 (2011). https://doi.org/10.1103/PhysRevLett.107.157201
7. Y. Tang, A.W. Sandvik, Phys. Rev. B **92**, 184425 (2015). https://doi.org/10.1103/PhysRevB.92.184425
8. J. Lou, A.W. Sandvik, N. Kawashima, Phys. Rev. B **80**, 180414 (2009). https://doi.org/10.1103/PhysRevB.80.180414
9. A.W. Sandvik, Phys. Rev. Lett. **104**, 177201 (2010). https://doi.org/10.1103/PhysRevLett.104.177201
10. S. Jin, A.W. Sandvik, Phys. Rev. B **87**, 180404 (2013). https://doi.org/10.1103/PhysRevB.87.180404
11. Y. Tang, A.W. Sandvik, Phys. Rev. Lett. **110**, 217213 (2013). https://doi.org/10.1103/PhysRevLett.110.217213
12. A.W. Sandvik, J. Kurkijärvi, Phys. Rev. B **43**, 5950 (1991). https://doi.org/10.1103/PhysRevB.43.5950
13. O.F. Syljuåsen, A.W. Sandvik, Phys. Rev. E **66**, 046701 (2002). https://doi.org/10.1103/PhysRevE.66.046701
14. K. Hukushima, K. Nemoto, J. Phys. Soc. Jpn. **65**, 1604 (1996). https://doi.org/10.1143/JPSJ.65.1604
15. P. Sengupta, A.W. Sandvik, D.K. Campbell, Phys. Rev. B **65**, 155113 (2002). https://doi.org/10.1103/PhysRevB.65.155113
16. I.S. Jacobs, P.E. Lawrence, Phys. Rev. **164**, 866 (1967). https://doi.org/10.1103/PhysRev.164.866
17. E. Stryjewski, N. Giordano, Adv. Phys. **26**, 487 (1977). https://doi.org/10.1080/00018737700101433
18. A.A. Aligia, Phys. Rev. B **63**, 014402 (2000). https://doi.org/10.1103/PhysRevB.63.014402
19. M. Arlego, F. Heidrich-Meisner, A. Honecker, G. Rossini, T. Vekua, Phys. Rev. B **84**, 224409 (2011). https://doi.org/10.1103/PhysRevB.84.224409
20. C. Gerhardt, K.-H. Mütter, H. Kröger, Phys. Rev. B **57**, 11504 (1998). https://doi.org/10.1103/PhysRevB.57.11504
21. S. Hirata, ArXiv e-prints (1999). cond-mat/9912066
22. D.V. Dmitriev, V.Y. Krivnov, Phys. Rev. B **73**, 024402 (2006). https://doi.org/10.1103/PhysRevB.73.024402
23. L. Kecke, T. Momoi, A. Furusaki, Phys. Rev. B **76**, 060407 (2007). https://doi.org/10.1103/PhysRevB.76.060407
24. J. Sudan, A. Lüscher, A.M. Läuchli, Phys. Rev. B **80**, 140402 (2009). https://doi.org/10.1103/PhysRevB.80.140402
25. A.K. Kolezhuk, F. Heidrich-Meisner, S. Greschner, T. Vekua, Phys. Rev. B **85**, 064420 (2012). https://doi.org/10.1103/PhysRevB.85.064420
26. D.P. Landau, Phys. Rev. Lett. **28**, 449 (1972). https://doi.org/10.1103/PhysRevLett.28.449

27. P.A. Rikvold, W. Kinzel, J.D. Gunton, K. Kaski, Phys. Rev. B **28**, 2686 (1983). https://doi.org/10.1103/PhysRevB.28.2686
28. D. Huerga, J. Dukelsky, N. Laflorencie, G. Ortiz, Phys. Rev. B **89**, 094401 (2014). https://doi.org/10.1103/PhysRevB.89.094401
29. C.K. Majumdar, D.K. Ghosh, J. Math. Phys. **10**, 1388 (1969). https://doi.org/10.1063/1.1664978
30. C.K. Majumdar, D.K. Ghosh, J. Math. Phys. **10**, 1399 (1969). https://doi.org/10.1063/1.1664979
31. Z.G. Soos, A. Parvej, M. Kumar, J. Phys. Condens. Matter **28**, 175603 (2016). http://stacks.iop.org/0953-8984/28/i=17/a=175603
32. S. Sachdev, T. Senthil, R. Shankar, Phys. Rev. B **50**, 258 (1994). https://doi.org/10.1103/PhysRevB.50.258
33. A. Iaizzi, A.W. Sandvik, J. Phys. Conf. Ser. **640**, 012043 (2015). http://stacks.iop.org/1742-6596/640/i=1/a=012043
34. T. Sulejmanpasic, H. Shao, A.W. Sandvik, M. Ünsal, Phys. Rev. Lett. **119**, 091601 (2017). https://doi.org/10.1103/PhysRevLett.119.091601
35. F.D.M. Haldane, Phys. Rev. B **25**, 4925 (1982). https://doi.org/10.1103/PhysRevB.25.4925
36. C.K. Majumdar, J. Phys. C Solid State Phys. **3**, 911 (1970). http://stacks.iop.org/0022-3719/3/i=4/a=019
37. R.R.P. Singh, M.E. Fisher, R. Shankar, Phys. Rev. B **39**, 2562 (1989). https://doi.org/10.1103/PhysRevB.39.2562
38. M. Mourigal, M. Enderie, A. Klöpperpieper, J.S. Caux, A. Stunault, H. Rønnow, Nat. Phys. **9**, 435 (2013). https://doi.org/10.1038/nphys2652
39. F. Heidrich-Meisner, A. Honecker, T. Vekua, Phys. Rev. B **74**, 020403 (2006). https://doi.org/10.1103/PhysRevB.74.020403
40. J. Schnack, H.-J. Schmidt, J. Richter, J. Schulenburg, Eur. Phys. J. B **24**, 475 (2001). https://doi.org/10.1007/s10051-001-8701-6
41. J. Schulenburg, A. Honecker, J. Schnack, J. Richter, H.-J. Schmidt, Phys. Rev. Lett. **88**, 167207 (2002). https://doi.org/10.1103/PhysRevLett.88.167207
42. J. Richter, J. Schulenburg, A. Honecker, J. Schnack, H.-J. Schmidt, J. Phys. Condens. Matter **16**, S779 (2004). http://stacks.iop.org/0953-8984/16/i=11/a=029
43. A. Honecker, J. Schulenburg, J. Richter, J. Phys. Condens. Matter **16**, S749 (2004). http://stacks.iop.org/0953-8984/16/i=11/a=025
44. L. Balents O.A. Starykh, Phys. Rev. Lett. **116**, 177201 (2016). https://doi.org/10.1103/PhysRevLett.116.177201
45. M. Oshikawa, M. Yamanaka, I. Affleck, Phys. Rev. Lett. **78**, 1984 (1997). https://doi.org/10.1103/PhysRevLett.78.1984
46. F.D.M. Haldane, Phys. Rev. Lett. **47**, 1840 (1981). https://doi.org/10.1103/PhysRevLett.47.1840
47. T. Tonegawa, I. Harada, J. Phys. Soc. Jpn. **56**, 2153 (1987). https://doi.org/10.1143/JPSJ.56.2153
48. D.J.J. Farnell, J.B. Parkinson, J. Phys. Condens. Matter **6**, 5521 (1994). http://stacks.iop.org/0953-8984/6/i=28/a=024
49. C. Gerhardt, A. Fledderjohann, E. Aysal, K.-H. Mütter, J.F. Audet, H. Kröger, J. Phys. Condens. Matter **9**, 3435 (1997). http://stacks.iop.org/0953-8984/9/i=16/a=015
50. H.T. Lu, Y.J. Wang, S. Qin, T. Xiang, Phys. Rev. B **74**, 134425 (2006). http://dx.doi.org/10.1103/PhysRevB.74.134425
51. F. He, Y. Jiang, Y.-C. Yu, H.-Q. Lin, X.-W. Guan, Phys. Rev. B **96**, 220401 (2017). http://dx.doi.org/10.1103/PhysRevB.96.220401
52. Y. Kono, T. Sakakibara, C.P. Aoyama, C. Hotta, M.M. Turnbull, C.P. Landee, Y. Takano, Phys. Rev. Lett. **114**, 037202 (2015). http://dx.doi.org/10.1103/PhysRevLett.114.037202
53. M. Jeong, H.M. Rønnow, Phys. Rev. B **92**, 180409 (2015). http://dx.doi.org/10.1103/PhysRevB.92.180409

54. B.C. Watson, V.N. Kotov, M.W. Meisel, D.W. Hall, G. E. Granroth, W.T. Montfrooij, S.E. Nagler, D.A. Jensen, R. Backov, M.A. Petruska, G.E. Fanucci, D.R. Talham, Phys. Rev. Lett. **86**, 5168 (2001). http://dx.doi.org/10.1103/PhysRevLett.86.5168
55. B.-B. Mao, C. Cheng, F.-Z. Chen, H.-G. Luo, Sci. Rep. **7**, 18104 (2017). http://dx.doi.org/10.1038/s41598-017-17887-w
56. H. Suwa, *Geometrically Constructed Markov Chain Monte Carlo Study of Quantum Spin-Phonon Complex Systems* (Springer Japan, Tokyo, 2014), Chap. 6. http://dx.doi.org/10.1007/978-4-431-54517-0
57. A.W. Sandvik, Phys. Rev. B **59**, R14157 (1999). http://dx.doi.org/10.1103/PhysRevB.59.R14157

Chapter 3
Saturation Transition in the 2D J-Q Model

We present a study of metamagnetism and zero-scale-factor universality at the saturation transition of the $S = \frac{1}{2}$ J-Q model in the presence of an external (Zeeman) magnetic field—the J-Q-h model. Metamagnetism is a kind of first-order phase transition in which the magnetization changes discontinuously as a function of field. This kind of transition usually occurs in systems with frustrated or anisotropic spin interactions, but recent work [2] has shown that metamagnetism can occur in the one-dimensional (1D) J-Q model, which lacks these properties. The J-Q model is a quantum antiferromagnet formed from the Heisenberg exchange (J) augmented with a four-spin interaction (Q) of the form $-QP_{i,j}P_{k,l}$ (where $P_{i,j} \equiv \frac{1}{4} - \mathbf{S}_i \cdot \mathbf{S}_j$). In our previous work on the one-dimensional J-Q model, we found that the transition to saturation is first order (i.e., metamagnetic) above a critical coupling ratio $(Q/J)_{\text{min}}$ [2, 3],[1] below which the saturation transition is continuous and governed by a zero-scale-factor universal critical scaling form [2, 4]. We present a study of the two-dimensional (2D) J-Q model. As in the 1D case, we find a metamagnetic transition to saturation driven by an identical mechanism and report an exact solution for the critical coupling ratio $(Q/J)_{\text{min}}$ where the jump first appears. We also study the saturation transition in the continuous limit, which we expect to be governed by zero-scale-factor universality at its upper critical dimension [2, 4]. This is the first numerical study of the zero-scale-factor universality in two dimensions. We find multiplicative logarithmic violations of the universal scaling relation which do not match the form proposed by Sachdev *et al.* [4] and discuss an alternative form based on the 4D Ising universality class.

A version of this chapter without the discussion on 4D Ising universality titled "Metamagnetism and zero-scale-factor universality in the two-dimensional J-Q model" and coauthored with Anders W. Sandvik and Kedar Damle has been published in Physical Review B **98** 064405 (2018) [1]. Reprinted with permission.

[1] See also, Chap. 2.

A. Iaizzi, *Magnetic Field Effects in Low-Dimensional Quantum Magnets*,
Springer Theses, https://doi.org/10.1007/978-3-030-01803-0_3

3.1 Background

The J-Q model is part of a family of Marshall positive[2] Hamiltonians constructed from products of singlet projection operators [5]. The two-dimensional realization of the J-Q model is given by

$$H_{JQ} = -J \sum_{\langle i,j \rangle} P_{i,j} - Q \sum_{\langle i,j,k,l \rangle} P_{i,j} P_{k,l} \tag{3.1}$$

where $\langle i, j \rangle$ sums over nearest neighbors and $\langle i, j, k, l \rangle$ sums over plaquettes on a square lattice as pairs acting on rows $\substack{k\ l \\ i\ j}$ and columns $\substack{j\ l \\ i\ k}$ [6]. The zero-field J-Q model has been well studied in both one [2, 7–9] and two [6, 10–13] dimensions, where it provides a numerically tractable way to study the deconfined quantum critical point marking the transition between the Néel antiferromagnetic state and the valence-bond solid (VBS). In the VBS spins break Z_4 lattice symmetry to form singlet pairs with their neighbors in an ordered fashion (see Fig. 4.1). Here we will not focus on this aspect of the J-Q model, but instead add an external magnetic field h_z

$$H_{JQh} = H_{JQ} - h_z \sum_i S_i^z, \tag{3.2}$$

and study the magnetization near the field-driven transition to saturation. Hereafter we will either fix the energy scale by setting $J = 1$ (and referring to the dimensionless parameters $q \equiv Q/J$ and $h \equiv h_z/J$) or by requiring $J + Q = 1$ (and referring to the dimensionless parameters $s \equiv Q/(J + Q)$ and $h \equiv h_z/(J + Q)$).

Metamagnetism (or magnetization jumps) is a first-order phase transition in which the magnetization changes suddenly in response to an infinitesimal change in the magnetic field [14, 15]. This sort of transition usually occurs in spin systems with frustration or intrinsic anisotropy [16–24]. A 1D version of the J-Q model (studied in Chap. 2) undergoes magnetization jumps to saturation above a critical coupling ratio $q_{\min} = 2/9$ caused by the onset of attractive interactions between magnons (flipped spins against a fully polarized background) mediated by the four-spin interaction [2]. In the 1D case the critical coupling ratio $q_{\min}$ can be determined exactly using a high-magnetization expansion [2]. Here we generalize previous work to include the 2D case, we find metamagnetism caused by the same mechanism and determine $q_{\min}$ to numerical precision using an exact method.

Zero-scale-factor universality occurs when the response functions depend on the bare coupling constants and no nonuniversal numbers [4]. It applies to continuous quantum phase transitions that feature the onset of a nonzero ground state

[2]The term "Marshall positive" refers to Hamiltonians that are free of the sign problem and therefore accessible to large-scale numerical study by quantum Monte Carlo simulations. See Sect. 5.3.1 for an explanation of the sign problem.

expectation value of a conserved density [2, 4]. The saturation transition in the J-Q model for $q < q_{\min}$ is just such a situation [2], although the 2D case is at the upper critical dimension of the theory, so we expect to find multiplicative logarithmic corrections to the universal scaling form.

Outline Outline the methods used in this chapter are summarized in Sect. 3.2; in Sect. 3.3, we discuss a schematic phase diagram of the 2D J-Q model; in Sect. 3.4, we discuss the magnetization jump and derive an exact solution for $q_{\min}$ (where the saturation transition becomes first order); in Sect. 3.5 we discuss the universal scaling of the saturation in the continuous regime at the upper critical dimension of the zero-scale-factor universality and logarithmic violations of the scaling form; conclusions are discussed in Sect. 3.6.

3.2 Methods

For the exact solution for $q_{\min}$ we have used Lanczos exact diagonalization [25]. The large-scale numerical results in this chapter were generated using the stochastic series expansion quantum Monte Carlo (QMC) method with directed loop updates [26] and quantum replica exchange based on the method used in our previous work [2]. Stochastic series expansion is a QMC method which maps a d-dimensional quantum problem onto a $(d + 1)$-dimensional classical problem by means of a Taylor expansion of the density matrix $\rho = e^{-\beta H}$, where the extra dimension roughly corresponds to imaginary time in a path-integral formulation [25]. In the QMC sampling, the emphasis is on the operators that move the world-lines rather than the lines themselves. The method used here is based originally on the method described in [26]. In addition to the standard updates, we incorporated quantum replica exchange [27, 28], a multicanonical method in which the magnetic field (or some other parameter) is sampled stochastically by running many simulations in parallel with different magnetic fields and periodically allowing them to swap fields in a manner that obeys the detailed balance condition. A detailed description of these techniques can be found in Chap. 5.

3.3 Phase Diagram

In Fig. 3.1, we present a schematic zero-temperature phase diagram of the 2D J-Q model combining previous work with the results presented in this chapter. The h-axis of Fig. 3.1 corresponds to the well-understood 2D Heisenberg antiferromagnet in an external field, and the q-axis corresponds to the previously studied [6, 10–13, 29] zero-field J-Q model, which for $q < q_c$ has long-range antiferromagnetic Néel order in the ground state. At finite temperature $O(3)$ spin-rotation symmetry (which is continuous) cannot be spontaneously broken

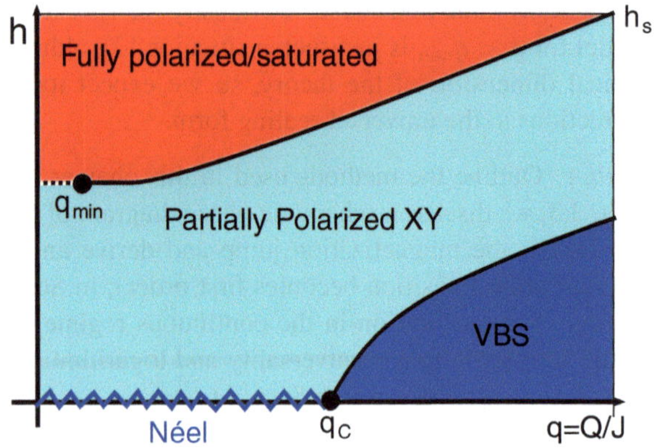

Fig. 3.1 Cartoon phase diagram of the 2D J-Q model in an external field at zero temperature. The different phases and critical points are explained in the text

(according to the Mermin–Wagner Theorem [30]), so there is no long-range spin order; instead there is a "renormalized classical" regime with the spin correlation length diverging exponentially as $T \to 0$ like $\xi \propto e^{2\pi\rho_s/T}$ [31]. At q_c, the zero-field J-Q model undergoes a quantum phase transition to the valence-bond solid (VBS) state. The off-axes area of Fig. 3.1 has not previously been studied; we here focus on the region around the field-driven saturation transition. The region around the deconfined quantum critical point, q_c, is addressed in Chap. 4.

Starting from the Néel state ($q < q_c$), adding a magnetic field forces the XY correlations into the XY plane, producing a partially polarized canted antiferromagnetic state. At finite temperature, there is no long-range Néel order, but the addition of a field permits a BKT-like transition to a phase with power-law spin correlations. For $q > q_c$, the ground state has VBS order. This state has a finite gap, so it survives at finite temperature and is destroyed by the magnetic field only after it closes the spin gap. The destruction of the VBS recovers the canted antiferromagnetic state (or partially polarized spin disordered phase for $T > 0$). We discuss these transitions in more detail in Chap. 4

We here will focus on the saturation transition in the high-field region of the phase diagram. The system reaches saturation (where all spins are uniformly aligned in the $+z$ direction) at $h = h_s(q)$. For $q < q_{\min}$, this transition is continuous and the saturation field is given by $h_s = 4J$. At the tricritical point, $q_{\min}$, the magnetic susceptibility diverges at saturation (corresponding to an infinite-order phase transition). For $q > q_{\min}$ is further increased, the transition to saturation is first order: a macroscopic jump in the magnetization known as the metamagnetism.

3.4 Metamagnetism

Magnetization jumps (also known as metamagnetism) can appear due to a variety of mechanisms including broken lattice symmetries, magnetization plateaus [32], localization of magnetic excitations [33–35], and bound states of magnons

[2, 18, 20]. It has previously been established that magnetization jumps occur in the J-Q chain caused by the onset of a bound state of magnons [2, 3, 36]; this is the first known example of metamagnetism in the absence of frustration or intrinsic anisotropy. To understand the mechanism for metamagnetism, we consider bosonic spin flips (magnons) on a fully polarized background (see Chap. 2 for a full explanation of this approach). These magnons are hardcore bosons that interact with a short-range repulsive interaction in the Heisenberg limit. The introduction of the Q-term produces an effective short-range *attractive* interaction between magnons. At $q_{\min}$, this attractive force dominates and causes pairs of magnons to form bound states.

3.4.1 Exact Solution for $q_{\min}$

We will now find $q_{\min}$ for the 2D J-Q model using a similar procedure to the one used for the J-Q chain in Chap. 2. Let us define bare energy of an n-magnon state, $\bar{E}_n$, as

$$E_n(J, Q, h) = \bar{E}_n(J, Q) - nh/2. \tag{3.3}$$

We can then define the binding energy of two magnons as

$$\Xi(q) \equiv 2\bar{E}_1 - \bar{E}_2. \tag{3.4}$$

The single-magnon energy, $\bar{E}_1$, can be analytically determined to be $\bar{E}_1 = -4J$. The two-magnon energy, $\bar{E}_2$, must be determined numerically, but since this is a two-body problem, relatively large systems can be studied using Lanczos exact diagonalization.

In Fig. 3.2 we plot the binding energy of two magnons, $\Xi(q, L)$, for $0 \leq q \leq 1$ and $L = 4,\ 8,\ 12,\ 16$. For all sizes the binding energy becomes positive around $q \approx 0.417$. We can also see that Fig. 3.2 strongly resembles the analogous figure for the J-Q chain (see Fig. 2.6). For $q < q_{\min}$ finite-size effects result in an *underestimate* of the binding energy and for $q > q_{\min}$ finite-size effects cause an *overestimate* of the binding energy. Around $q_{\min}$ these effects cancel out and the crossing is *nearly* independent of system size (in the 1D case the crossing is exactly independent of L). Using a bracketing procedure, we can extract $q_{\min}(L)$ to numerical precision. Table 3.1 contains a list of $q_{\min}(L)$ for select $L \times L$ systems with $L \leq 24$. $q_{\min}$ converges exponentially fast in L, so even based on these modest sizes we know $q_{\min}(L = \infty) = 0.41748329$ to eight digits of precision. Although we do not plot it here, the exponential convergence of $q_{\min}(L)$ can be seen from the underlines in Table 3.1, which indicate the digits which are converged to the thermodynamic limit; the number of underlined digits grows linearly with L. Note here that $q_{\min}$ is not the same as q_c (the Néel-VBS transition point), and these two phase transitions are governed by completely different physics.

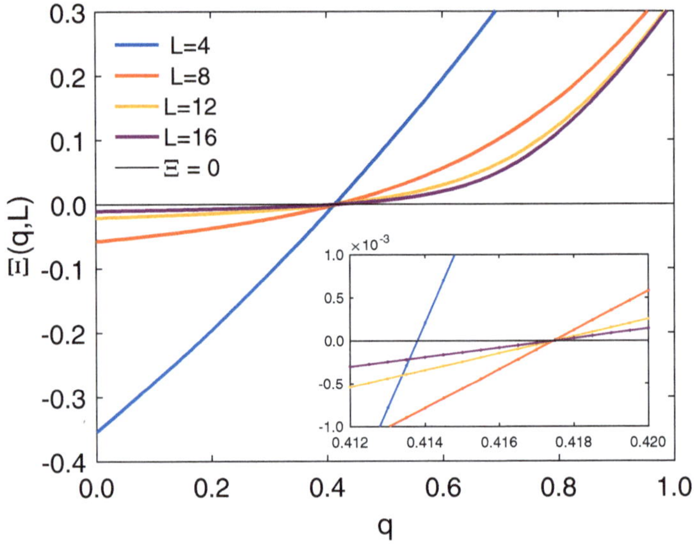

Fig. 3.2 Binding energy $\Xi(q, L)$ plotted against q for several systems sizes calculated using exact diagonalization. The thin black line represents $\Xi = 0$. Inset: zoomed-in view of crossing point

Table 3.1 $q_{\min}(L)$ calculated to machine precision for select $L \times L$ systems using Lanczos exact diagonalization. The underlined portions of the numbers represent the digits that are fully converged to the thermodynamic limit

L	$q_{\min}$
4	0.413793103448
6	0.417287630402
8	0.417467568061
10	0.417481179858
12	0.417482857341
14	0.417483171909
16	0.417483250752
18	0.417483274856
20	0.417483283375
22	0.417483286742
24	0.417483288198

In Fig. 3.3 we plot the ground state probability density as a function of separation of the magnons in the x-direction, r_x (with $r_y = 0$). Here we consider a small (18×18) system in order to make the features at the boundary easier to distinguish on the scale of the figure. For $q = 0$, we can see that the probability density takes on the form of a free particle with periodic boundary conditions in r_x, r_y, with a single excluded site at $r_x = r_y = 0$. In the continuum limit, this corresponds to a repulsive delta potential. For $q > q_{\min}$ the wavefunction takes on the exponentially-decaying form of a bound state. At $q = q_{\min}$, the crossover between repulsive and attractive interactions, the wavefunction becomes flat with an exponentially-decaying short-distance disturbance of the form $\psi \propto 1 - ae^{r_x/b}$ (this was confirmed by further data

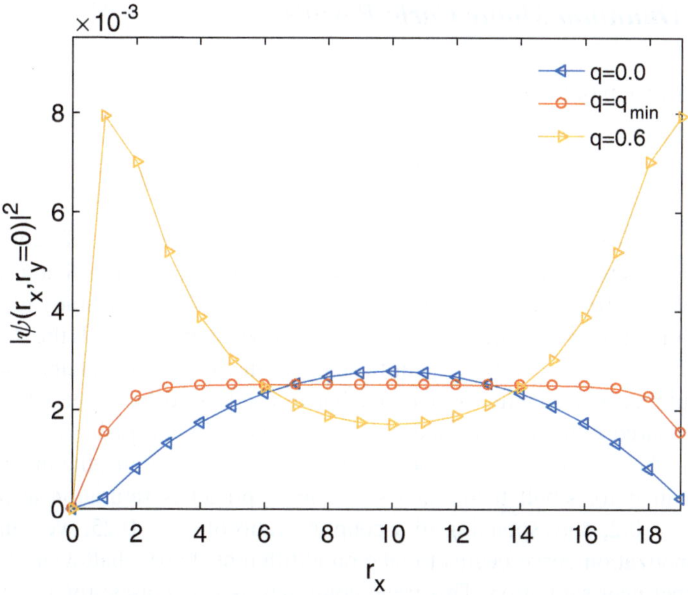

Fig. 3.3 Probability density of magnon separation in the x-direction for $r_y = 0$, $|\psi(r_x, r_y = 0)|$ in the two-magnon sector of the J-Q model; calculated using Lanczos exact diagonalization

not depicted here). This exponential disturbance explains why the finite-size effects vanish exponentially near $q_{\min}$. The wavefunction in the 2D case stands in contrast to the flat wavefunction in the 1D J-Q model, where the bulk wavefunction at $q_{\min}$ is perfectly flat and $q_{\min}$ is exactly independent of L for $L > 6$ [2].

The onset of attractive interactions between magnons has previously been found to cause metamagnetism [2, 18, 20]. The existence of a bound state of two magnons, as we have found here, is not a sufficient condition to guarantee the existence of a macroscopic magnetization jump. The magnetization could, for example, change by steps of $\Delta m_z = 2$, but never achieve a macroscopic jump [20, 37]. For a true jump to occur, the point $q_{\min}$ must be the beginning of an instability leading to ever larger bound states of magnons. In the next section we will confirm that a macroscopic magnetization jump does in fact occur using full magnetization curves generated by quantum Monte Carlo simulations. It will not be possible to detect the onset of the magnetization jump (which is initially infinitesimal) by directly examining the magnetization curves due to finite-temperature rounding. Instead in Sect. 3.5 we will examine the scaling of the magnetization near saturation and find that a qualitative change in behavior consistent with the onset of a different universality occurs at the predicted value of $q_{\min}$.

3.4.2 Quantum Monte Carlo Results

In Fig. 3.4, we plot the magnetization density,

$$m = \frac{2}{L^2} \sum S_i^z, \tag{3.5}$$

of the 2D J-Q model as a function of external field for several different values of $0 \leq s \leq 1$ where s is defined such that $J = 1 - s$ and $Q = s$ such that $J + Q = 1$. Here we use a 16×16 lattice with $\beta = 4$. Ordinarily, QMC can study much larger systems than this, but as was observed in our previous work [2, 3], the J-Q model with a field is exceptionally difficult to study, even when using enhancements such as β-doubling and quantum replica exchange (both used here, see Chap. 5). We have compared to larger sizes and finite-size effects do not qualitatively effect the results on the scale of Fig. 3.4. For $s = 0$ (the Heisenberg limit), the magnetization is linear in h for small fields, and smoothly approaches saturation at $h = 4J$. When $s = 0.2$, corresponding to a coupling ratio of $q = 0.25$, we can see that the magnetization curve begins to take on a different shape: shallower at low field and steeper near saturation. This trend continues as s increases: for $s \geq 0.8$, there is a clear discontinuity in the magnetization. Although the jump should appear for $q \geq q_{\text{min}} = 0.417$, which corresponds to $s_{\text{min}} = 0.294$, this is difficult to distinguish in the QMC data. At q_{min}, the jump is infinitesimal, and even when the jump is

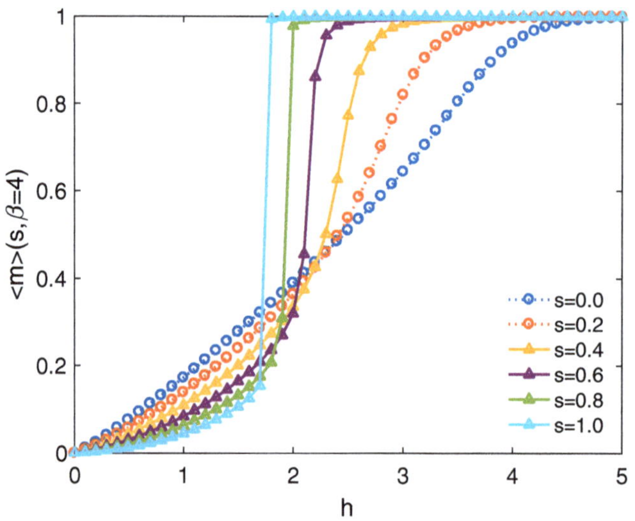

Fig. 3.4 Magnetization density of the 2D J-Q model as function of external field, h, for a range of different values of s defined such that $J = 1 - s$ and $Q = s$. Here $s = 0, 0.2, 0.4, 0.6, 0.8, 1$ with $\beta = 4$ correspond to $q = 0, 0.25, 0.67, 1.5, 4, \infty$, respectively (with rescaled non-constant β). Results from QMC with quantum replica exchange

larger, such as for $s = 0.4$ and $s = 0.6$, it is hard to clearly distinguish due to finite-temperature effects, which round off the discontinuity in the magnetization. These results are nonetheless consistent with the value of $q_{\min}$ predicted in Sect. 3.4.1, and demonstrate that a macroscopic magnetization jump does in fact occur. We will discuss more evidence for $q_{\min} \approx 0.417$ from the critical scaling of the magnetization near saturation in Sect. 3.5.

3.5 Zero-Scale-Factor Universality in 2D

In the J-Q model, magnetization near saturation should be governed by a remarkably simple zero-scale-factor universality for $q < q_{\min}$ (where the saturation transition is continuous) [2, 4]. Here, "zero-scale-factor" means that the response functions are universal functions of the bare coupling constants and do not depend on any nonuniversal numbers [4]. Zero-scale-factor universality applies to low-dimensional systems where there is a quantum phase transition characterized by a smooth onset of a conserved density [4]. Typically this is applied to the transition from the gapped singlet state of integer spin chains to a field-induced Bose–Einstein condensate of magnons (excitations above the zero magnetization state). In the J-Q model, we instead start from the saturated state with $h > h_s$, and consider flipped spins on this background—magnons—as h is decreased below h_s. In the 1D case, the zero-factor scaling form applies for all $q < q_{\min}$ for sufficiently low temperatures, and is violated by a logarithmic divergence at exactly $q_{\min}$ (see Sect. 2.6) [2]. The 2D J-Q model is at the upper critical dimension of this universality, so we expect to see multiplicative logarithmic violations of the zero-factor scaling form. We will describe the universal scaling form and its application to the saturation transition in the 2D J-Q model. We will also show that the low-temperature violation of the scaling form does not match the prediction in [4] and discuss other possible forms.

In two spatial dimensions, the zero-factor scaling form is given by Eq. (1.23) of [4]:

$$\langle m \rangle = g\mu_B \left(\frac{2M}{\hbar^2 \beta} \right) \mathcal{M}(\beta\mu) \tag{3.6}$$

where M is the magnon mass (which is $M = 1$ when $J = 1$), and μ represents the field, $\mu \equiv h_s - h$. For $q \leq q_{\min}$, the saturation field is $h_s = 4J$, which can be determined analytically from the level crossing between the saturated state and the state with a single flipped spin. We set $\hbar = 1$ and use the number density of magnons $\langle m \rangle = g\mu_B \langle n \rangle$ to define the rescaled magnon density:

$$n_s(q, \beta\mu) \equiv \frac{\beta \langle n \rangle}{2} = \mathcal{M}(\beta\mu) \tag{3.7}$$

In the limit $h \to h_s$, $\mu \to 0$, n_s will be independent of temperature. We emphasize again that these magnons are spin flips on fully polarized background, so $n \to 0$ corresponds to the saturated state. The field is also reversed from the usual case, where $h > h_s$ produces a negative μ, which means $n \to 0$, and $h < h_s$ corresponds to a positive μ and a finite density of magnons.

At the saturation field, $\mu = 0$, the scaling form in Eq. (3.7) predicts that the density takes on a simple form:

$$\langle n \rangle = 2\mathcal{M}(0)T \tag{3.8}$$

At this same point the rescaled density, n_s, becomes independent of temperature:

$$n_s(q, 0) \equiv \frac{\beta \langle n \rangle}{2} = \mathcal{M}(0) \tag{3.9}$$

However, in our case there are two spatial dimensions and $z = 2$ imaginary time dimensions, so the total dimensionality is $d = 4$. The upper critical dimension of the zero-scale-factor universality is $d = 4$ [4]. At low temperatures, we therefore expect to see multiplicative logarithmic violations of the scaling form Eq. (3.7).

In Fig. 3.5, we plot the rescaled magnon density at saturation, $n_s(q, \mu = 0)$, as a function of temperature for two different sizes, 32×32 and 64×64. Here we use the exact value of the saturation field $h_s(q \leq q_{\text{min}}) = 4J$. These sizes are large enough that finite-size effects only become important at low temperature; the results for the two different sizes overlap completely for $T \geq 0.1$, but exhibit some separation at lower temperature, depending on the value of q. If there were no corrections to the universal form, the lines in Fig. 3.5 would exhibit no temperature dependence. Instead, we observe violations of the scaling form for all q. For $q = 0$, there is some non-monotonic behavior, with a local minimum around $T = 0.35$; at low temperatures, $n_s(T)$ appears to diverge like $\log(1/T)$, which on this semi-log scale manifests as a straight line. For $q = 0.1$ and 0.2, the behavior is similar, although n_s has been shifted upwards. For $q = 0.3$, the local minimum in $n_s(T)$ appears to be gone. The divergence for $q < q_{\text{min}}$ looks log-linear, but it is difficult to distinguish between different powers of the log by fitting alone. As $q = 0.4$ and $q = q_{\text{min}} = 0.4174833$, finite-size effects become more important, and it is possible that the log has a different power. From simulations of 96×96 and 128×128 systems (not depicted here) we know that the 64×64 curve for $q = q_{\text{min}}$ is converged to the thermodynamic limit within error bars.

We can also use the low-temperature behavior of n_s in Fig. 3.5 to verify our prediction of q_{min} (from the high-magnetization expansion discussed in Sect. 3.4.1). At q_{min}, the transition is no longer the smooth onset of a conserved density, therefore the zero-scale-factor universality does not apply (not even with logarithmic corrections). For all $q < q_{\text{min}}$, the low-temperature divergence appears to obey a form $\log\left(\frac{1}{T}\right)$, or some power of it. The divergence of the $n_s(q_{\text{min}}, T)$ curve takes on a *qualitatively* different form that appears to diverge faster than $\log\left(\frac{1}{T}\right)$. This is a

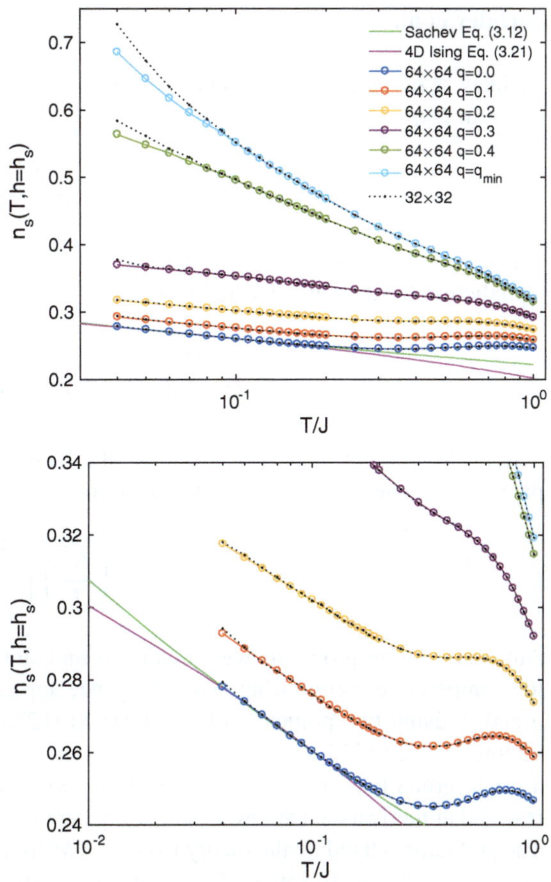

Fig. 3.5 (Top) The zero-factor-rescaled magnon density [Eq. (3.7)] at $h = h_s$, $\mu = 0$ calculated using QMC with quantum replica exchange. The bright green line is a fit to the scaling form Eq. (3.12), the magenta line is a fit to the 4D Ising scaling form Eq. (3.21). (Bottom) A zoomed-in view

confirmation that the predicted value of $q_{\min}$ is correct, even though no sign of a discontinuity can be observed in the magnetization curves themselves (Fig. 3.4) due to finite-temperature rounding.

3.5.1 Form of the Low-Temperature Divergence

We will now attempt to determine the form of the low-temperature divergence of n_s for $q < q_{\min}$, restricting ourselves to the Heisenberg limit, $q = 0$. We find that the divergence does not obey the form predicted in [4] for the zero-scale-factor universality at its upper critical dimension, and more closely matches the 4D Ising universality (also at its upper critical dimension).

Sachdev et al.

Sachdev *et al.* predict that at saturation ($h = h_s$), the magnon density will take on the form (Eq. (2.20) of [4]):

$$\langle n \rangle = \frac{2Mk_BT}{4\pi}\left[\log\left(\frac{\Lambda^2}{2Mk_BT}\right)\right]^{-4}. \tag{3.10}$$

where Λ is an upper (UV) momentum cutoff. We can plug this into Eq. (3.7) to find a prediction for the log-corrected form of $n_s(h = h_s)$:

$$n_s = \frac{M}{4\pi}\left[\log\left(\frac{\Lambda^2}{2Mk_BT}\right)\right]^{-4} \tag{3.11}$$

Setting the magnon mass, M, to unity (the bare value) and introducing a fitting parameter, a, we can fit $n_s(T \to 0)$ to the form:

$$n_s = a\left[\log\left(\frac{\Lambda^2}{T}\right)\right]^{-4}. \tag{3.12}$$

Automatic fitting programs were unable to find suitable values of a and Λ (in the low-temperature regime where the divergence appears), so we manually solved for a and Λ using two points: $n_s(T = 0.04) = 0.278$ and $n_s(T = 0.1) = 0.2604$, finding $a = 2.65354 \times 10^6$ and $\Lambda = 1.7 \times 10^{-13}$. We plot the resulting curve as a bright green line in Fig. 3.5. Although this *appears* to produce a good fit to the rescaled numerical data at low T, the fitting parameters do not make physical sense. The prefactor is fixed by the theory to be $a = M/(4\pi) \approx 0.08$, yet the fitted value is huge: $a \approx 10^6$. Worse yet, the UV cutoff, Λ, is extremely very small, much smaller than any other scale in this problem. In zero-scale-factor universality, there should be no renormalization of bare parameters, but even allowing for renormalization of M (due to being at the upper critical dimension), it is not possible for Eq. (3.12) to match the data while maintaining a physically sensible (i.e., large) value of the UV cutoff Λ.

The fit in Fig. 3.5 looks remarkably like a linear $\log T$ divergence. Indeed, since $T \gg \Lambda^2$, we can expand Eq. (3.12) in a Taylor series around small $u = \log T$ and we find an expression

$$n_s = \frac{a}{(\log \Lambda^2)^4}\left[1 + 4\frac{\log T}{\log \Lambda^2} + 10\left(\frac{\log T}{\log \Lambda^2}\right)^2 + \cdots\right] \tag{3.13}$$

that is linear in $\log T$ to first order and converges rapidly because $\log \Lambda^2 \approx -58$. Considering this fact and the unphysical parameters required to make the form fit, it is clear that Eq. (3.11) does not accurately predict the violations of the zero-scale-factor universality at its upper critical dimension. The apparent fit is instead a roundabout approximation of the true form, which is (approximately) proportional to log $\left(\frac{1}{T}\right)$ to some power. The reason why the form predicted in [4] fails is unclear at this time.

4D Ising Universality

Since the scaling form from [4] did not work, we consider a different form based on the scaling of the order parameter in the 4D Ising model (an $O(1)$ ϕ^4 theory). The order parameter of the transition we study here (the magnon density) is a scalar in (2 + 2) dimensions (for a total of four), matching the 4D Ising model (which is at the upper critical dimension of the Ising universality class). This correspondence is a little unusual, as the magnon density cannot be negative and there are no fluctuations (or entropy of any kind) when the density is zero. Further, in the saturation transition, there is an extra $U(1)$ symmetry not present in the 4D Ising model.[3] This extra $U(1)$ symmetry is not a symmetry of the order parameter, but it could allow for off-diagonal correlations that would compete with the order parameter. In fact, the off-diagonal order also vanishes as the system fully polarizes, therefore the two order parameters should couple in a nontrivial way. Nevertheless, given that the predictions from [4] do not seem to work, it is interesting to test an alternative scenario where the order parameters do not couple, in which case one may expect a simple 4D Ising transition of the magnetization density. We will now map this scaling of the finite-temperature, zero-field magnon density, $n(T > 0, \mu = 0)$, onto the finite-field, $T = T_c$ scaling of the 4D Ising order parameter in the thermodynamic limit: $m(T = T_c, h > 0)$. We will find that this universality produces a plausible match to the low-temperature violations of scaling we observe in the J-Q model near saturation (Fig. 3.5).

First we will show that the leading-order (non-log-corrected) scaling from the 4D Ising model matches the zero-scale-factor form that is known to work, Eq. (3.6). The leading-order scaling forms of the 4D Ising order parameter at $T = T_c$ and $h > 0$ in the thermodynamic limit are given by:

$$m_{L=\infty}(T = T_c, h) \propto h^{1/\delta} \tag{3.14a}$$

$$\xi_{L=\infty}(T = T_c, h) \propto h^{-\nu_c} \tag{3.14b}$$

(see Eqs. (1.9) and (1.12) on pp. 5 of [38]). Let us solve for $m(\beta)$ to leading order (without the log corrections). We invert Eq. (3.14b) to find the finite field h in terms of the correlation length ξ:

$$h \propto \xi^{-1/\nu_c} \tag{3.15}$$

[3]Thanks to Cenke Xu for pointing this out.

Table 3.2 Selected critical exponents for a d-dimensional $O(N)$ ϕ^4 theory [38, p. 32]

δ	ν_c	$\hat{\delta}$	$\hat{q}$
3	1/3	1/3	1/4

Now we will turn this infinite-size problem into a finite-size problem by replacing the correlation length ξ with the system size L.

$$h \propto L^{-1/\nu_c} \tag{3.16}$$

In this case, we are already converged in the two spatial dimensions, and decreasing T corresponds to increasing the ($z = 2$) imaginary time dimensions, so we can then replace L with the inverse temperature, β:

$$h \propto \beta^{-1/\nu_c} \tag{3.17}$$

and then plug this back into Eq. (3.14a) to get $m(\beta)$:

$$m \propto \beta^{-1/\delta\nu_c} \tag{3.18}$$

Now we have eliminated all references to finite field and converted this into a finite-temperature problem. From Table 3.2 we know that $\delta = 3$ and $\nu_c = 1/3$, so we get, to leading order:

$$m \propto T \tag{3.19}$$

which is consistent with the leading-order zero-factor scaling prediction [Eq. (3.6)] which, as we have already seen, matches our numerical results.

To determine the log-corrected form, we will repeat the leading-order derivation using the log-corrected scalings [4, p. 9]:

$$m_{L=\infty}(T = T_c, h) \propto h^{1/\delta} \, |\ln h|^{\hat{\delta}} \tag{3.20a}$$

$$\xi_{L=\infty}(T = T_c, h) \propto h^{-\nu_c} \, |\ln h|^{\hat{\nu}_c} \tag{3.20b}$$

Here the correlation length exceeds the systems size:

$$\xi_L(T = T_c) \propto L(\ln(L))^{\hat{q}} \tag{3.20c}$$

There is no way to invert Eq. (3.20b) to get $h(\xi)$ like we did in Eq. (3.15). Instead, we will ignore these intermediate log corrections (to $h(\xi)$ and $\xi(L)$), which corresponds to neglecting higher-order (and subleading) log corrections to the final form. This approximation works because the logs are slowly varying. Ignoring the intermediate log corrections, we plug the leading-order result from Eq. (3.17) directly into Eq. (3.20a):

$$m(T) \propto \beta^{-1/\delta\nu_c} \left|\ln \beta^{-1/\nu_c}\right|^{\hat{\delta}}$$

Using the log-corrected form for $\xi(L)$ would have produced a $\ln(\ln(T))$ contribution. Now to convert this proportionality into an equation, we add a prefactor a and a scale T_0, plug in $\hat{\delta} = 1/3$

$$n(T) = 2aT \left|\ln \frac{T_0}{T}\right|^{1/3}$$

and then plug the result into Eq. (3.7) to find an expression for the rescaled magnon density, n_s

$$n_s(T) = a \left|\ln \frac{T_0}{T}\right|^{1/3} \tag{3.21}$$

This form already looks like a qualitative match to the $(\log T)^p$ divergences we see in Fig. 3.5. Fitting to the quantum Monte Carlo results, Eq. (3.21) produces a good match to the low-temperature behavior of n_s for $a = 0.16$, $T_0 \approx 7.5$. We plot this fit as a magenta line in Fig. 3.5. Due to the nature of logs, it would be difficult to distinguish between the line fit produced by the 4D Ising universality [Eq. (3.21)] and the one from the Sachdev *et al.* form [Eq. (3.12)], based on the quality of the fit alone, but the 4D Ising form clearly makes more physical sense in terms of the parameters resulting from the fit. Though this observation of a good fit to the 4D Ising form is intriguing, the reason for the diagonal and off-diagonal order parameters to decouple would have to be explained before this scenario can be accepted. Further studies are needed to settle this issue.

3.6 Conclusions

Here we have presented a numerical study of the two-dimensional J-Q model in the presence of an external magnetic field, focusing on the field-induced saturation transition. Building on a previous version of this study which focused on the one-dimensional case [2, 3] (see also Chap. 2), we have found that for $q \geq q_{\min}$ there is metamagnetism (magnetization jumps) in the saturation transition. The 1D J-Q model [2] is the only previously known spin Hamiltonian to exhibit metamagnetism in the absence of frustration or intrinsic anisotropy. We have determined $q_{\min}$ to numerical precision using the same high-magnetization expansion discussed in [2] and confirmed $q_{\min}$ by observing a qualitative change in the low-temperature scaling behavior of the magnon density near saturation. This transition is caused by the onset of effectively attractive interactions between magnons (flipped spins against a fully polarized background) caused by the four-spin Q term. The same mechanism can explain the presence of metamagnetism in a similar ring exchange model [24]. In the regime of the continuous saturation transition $q < q_{\text{min}}$, the saturation transition is governed by a zero-scale-factor universality at its upper critical dimension [4]. This universality has already been shown apply to the 1D case [2]. We have presented the first-ever numerical test of the zero-scale-factor universality in two dimensions. We found that the low-temperature scaling violations do *not* obey the form proposed by [4], which predicts a divergence as a *negative* power of $\log T$ as $T \to 0$, and instead they appear to diverge as some *positive* power of $\log T$.

There are still some important unanswered questions here that can be addressed in the future. The correct form of the low-temperature violations of the zero-scale-factor universality is still unknown at this time. One that is established, it would be interesting to test that scaling form over the full range of its validity $0 \leq q < q_{\text{min}}$. We have not discussed the scaling of the magnetization density near saturation at the tricritical point q_{min}, where the zero-scale-factor universality does not apply, but this would also be a potentially interesting topic for future research. The behavior of this system at low fields has not been discussed here, but will be addressed in the next chapter.

References

1. A. Iaizzi, K. Damle, A.W. Sandvik, Phys. Rev. B **98**, 064405 (2018). http://dx.doi.org/10.1103/PhysRevB.98.064405
2. A. Iaizzi, K. Damle, A.W. Sandvik, Phys. Rev. B **95**, 174436 (2017). http://dx.doi.org/10.1103/PhysRevB.95.174436
3. A. Iaizzi A.W. Sandvik, J. Phys. Conf. Ser. **640**, 012043 (2015). http://stacks.iop.org/1742-6596/640/i=1/a=012043
4. S. Sachdev, T. Senthil, R. Shankar, Phys. Rev. B **50**, 258 (1994). http://dx.doi.org/10.1103/PhysRevB.50.258
5. R.K. Kaul, R.G. Melko, A.W. Sandvik, Annu. Rev. Condens. Matter Phys. **4**, 179 (2013). http://dx.doi.org/10.1146/annurev-conmatphys-030212-184215
6. A.W. Sandvik, Phys. Rev. Lett. **98**, 227202 (2007). http://dx.doi.org/10.1103/PhysRevLett.98.227202
7. S. Sanyal, A. Banerjee, K. Damle, Phys. Rev. B **84**, 235129 (2011). http://dx.doi.org/10.1103/PhysRevB.84.235129
8. Y. Tang, A.W. Sandvik, Phys. Rev. Lett. **107**, 157201 (2011). http://dx.doi.org/10.1103/PhysRevLett.107.157201
9. Y. Tang, A.W. Sandvik, Phys. Rev. B **92**, 184425 (2015). http://dx.doi.org/10.1103/PhysRevB.92.184425
10. J. Lou, A.W. Sandvik, N. Kawashima, Phys. Rev. B **80**, 180414 (2009). http://dx.doi.org/10.1103/PhysRevB.80.180414
11. A.W. Sandvik, Phys. Rev. Lett. **104**, 177201 (2010). http://dx.doi.org/10.1103/PhysRevLett.104.177201
12. S. Jin, A.W. Sandvik, Phys. Rev. B **87**, 180404 (2013). http://dx.doi.org/10.1103/PhysRevB.87.180404
13. Y. Tang, A.W. Sandvik, Phys. Rev. Lett. **110**, 217213 (2013). http://dx.doi.org/10.1103/PhysRevLett.110.217213

14. I.S. Jacobs, P.E. Lawrence, Phys. Rev. **164**, 866 (1967). http://dx.doi.org/10.1103/PhysRev.164.866
15. E. Stryjewski, N. Giordano, Adv. Phys. **26**, 487 (1977). http://dx.doi.org/10.1080/00018737700101433
16. C. Gerhardt, K.-H. Mütter, H. Kröger, Phys. Rev. B **57**, 11504 (1998). http://dx.doi.org/10.1103/PhysRevB.57.11504
17. S. Hirata, ArXiv e-prints (1999). http://arxiv.org/abs/cond-mat/9912066cond-mat/9912066
18. A.A. Aligia, Phys. Rev. B **63**, 014402 (2000). http://dx.doi.org/10.1103/PhysRevB.63.014402
19. D.V. Dmitriev, V.Y. Krivnov, Phys. Rev. B **73**, 024402 (2006). http://dx.doi.org/10.1103/PhysRevB.73.024402
20. L. Kecke, T. Momoi, A. Furusaki, Phys. Rev. B **76**, 060407 (2007). http://dx.doi.org/10.1103/PhysRevB.76.060407
21. J. Sudan, A. Lüscher, A.M. Läuchli, Phys. Rev. B **80**, 140402 (2009). http://dx.doi.org/10.1103/PhysRevB.80.140402
22. M. Arlego, F. Heidrich-Meisner, A. Honecker, G. Rossini, T. Vekua, Phys. Rev. B **84**, 224409 (2011). http://dx.doi.org/10.1103/PhysRevB.84.224409
23. A.K. Kolezhuk, F. Heidrich-Meisner, S. Greschner, T. Vekua, Phys. Rev. B **85**, 064420 (2012). http://dx.doi.org/10.1103/PhysRevB.85.064420
24. D. Huerga, J. Dukelsky, N. Laflorencie, G. Ortiz, Phys. Rev. B **89**, 094401 (2014). http://dx.doi.org/10.1103/PhysRevB.89.094401
25. A.W. Sandvik, in *American Institute of Physics Conference Series* ed. by A. Avella, F. Mancini, vol. 1297, (2010), pp. 135–338, http://arxiv.org/abs/1101.3281. http://dx.doi.org/10.1063/1.3518900
26. O.F. Syljuåsen, A.W. Sandvik, Phys. Rev. E **66**, 046701 (2002). http://dx.doi.org/10.1103/PhysRevE.66.046701
27. K. Hukushima, K. Nemoto, J. Phys. Soc. Jpn. **65**, 1604 (1996). http://dx.doi.org/10.1143/JPSJ.65.1604
28. P. Sengupta, A.W. Sandvik, D.K. Campbell, Phys. Rev. B **65**, 155113 (2002). http://dx.doi.org/10.1103/PhysRevB.65.155113
29. H. Shao, W. Guo, A.W. Sandvik, Science **352**, 213 (2016). http://dx.doi.org/10.1126/science.aad5007
30. N.D. Mermin, H. Wagner, Phys. Rev. Lett. **17**, 1133 (1966). http://dx.doi.org/10.1103/PhysRevLett.17.1133
31. S. Chakravarty, B.I. Halperin, D.R. Nelson, Phys. Rev. Lett. **60**, 1057 (1988). http://dx.doi.org/10.1103/PhysRevLett.60.1057
32. A. Honecker, J. Schulenburg, J. Richter, J. Phys. Condens. Matter **16**, S749 (2004). http://stacks.iop.org/0953-8984/16/i=11/a=025
33. J. Richter, J. Schulenburg, A. Honecker, J. Schnack, H.-J. Schmidt, J. Phys. Condens. Matter **16**, S779 (2004). http://stacks.iop.org/0953-8984/16/i=11/a=029
34. J. Schnack, H.-J. Schmidt, J. Richter, J. Schulenburg, Eur. Phys. J. B **24**, 475 (2001). http://dx.doi.org/10.1007/s10051-001-8701-6
35. J. Schulenburg, A. Honecker, J. Schnack, J. Richter, H.-J. Schmidt, Phys. Rev. Lett. **88**, 167207 (2002). http://dx.doi.org/10.1103/PhysRevLett.88.167207
36. B.-B. Mao, C. Cheng, F.-Z. Chen, H.-G. Luo, Sci. Rep. **7**, 18104 (2017). http://dx.doi.org/10.1038/s41598-017-17887-w
37. F. Heidrich-Meisner, A. Honecker, T. Vekua, Phys. Rev. B **74**, 020403 (2006). http://dx.doi.org/10.1103/PhysRevB.74.020403
38. R. Kenna, Universal Scaling Relations for Logarithmic-Correction Exponents, in *Order, Disorder and Criticality: Advanced Problems of Phase Transition Theory*, vol. 3 (World Scientific Publishing Co, 2013), pp. 1–46. http://dx.doi.org/10.1142/9789814417891_0001

Chapter 4
Signatures of Deconfined Quantum Criticality in the 2D J-Q-h Model

Deconfined quantum criticality (DQC) is a type of quantum critical behavior characterized by the presence of exotic fractionalized excitations [1]. The transition between the Néel antiferromagnet and the valence-bond solid in the J-Q model is believed [2–7] to be an example of such a deconfined quantum critical point [1] where the excitations are spinons—bosons carrying spin-$1/2$ that correspond to vortices of the valence-bond solid order parameter [8, 9]. Using a magnetic field, one can induce a finite ground state density of magnetic excitations, which can then form a Bose–Einstein condensate (BEC) at low temperature. Previous work [10] has predicted that a BEC of spinons would have an anomalous temperature dependence due to the presence of a gapless quadratic mode, providing a way to distinguish deconfined spinons from conventional (nonfractional) magnons. In this preliminary report, we study this possibility using quantum Monte Carlo methods. We do not find the predicted anomalous temperature dependence in the spinon BEC, possibly due to the influence of an ignored dynamical gauge field [11]. At higher temperatures, the spinons are in a gaseous rather than BEC phase; this gaseous phase should also have a gapless quadratic mode [11]. In this deconfined spinon gas state we *do* detect the expected anomalous temperature dependence, providing direct evidence for the existence of deconfined spinons.[1] The introduction of a field also admits a phase transition of the Berezinskii–Kosterlitz–Thouless (BKT) type [12], which separates the spinon BEC and spinon gas phase. We estimate $T_{\mathrm{BKT}}(h)$ and also show that this transition produces a non-monotonic temperature dependence in the magnetization.

[1]In a forthcoming paper [13] in collaboration with the authors of [10], we will discuss thermodynamic behavior of both the spinon BEC and spinon gas accounting for the $U(1)$ gauge field and finite-temperature renormalization of the spinon mass.

A. Iaizzi, *Magnetic Field Effects in Low-Dimensional Quantum Magnets*,
Springer Theses, https://doi.org/10.1007/978-3-030-01803-0_4

4.1 Background

4.1.1 The Zero-Field J-Q Model

The 2D J-Q_2 model is a "designer Hamiltonian" custom-built for studying deconfined quantum criticality. It augments the antiferromagnetic Heisenberg exchange with a product of two singlet projection operators [2]:

$$P_{i,j} = \frac{1}{4} - \vec{S}_i \cdot \vec{S}_j \tag{4.1}$$

Written in terms of these singlet projection operators, the J-Q model is given by:

$$H_{JQ} = -J \sum_{\langle i,j \rangle} P_{i,j} - Q \sum_{\langle i,j,k,l \rangle} P_{i,j} P_{k,l} \tag{4.2}$$

where $\langle i, j \rangle$ represents a sum over nearest neighbors and $\langle i, j, k, l \rangle$ represents a sum over plaquettes with "bonds" i, j and k, l arranged as parallel links in the horizontal $\begin{smallmatrix} k & l \\ i & j \end{smallmatrix}$ and vertical $\begin{smallmatrix} j & l \\ i & k \end{smallmatrix}$ directions (on a square lattice). In the Heisenberg limit, its ground state is a Néel antiferromagnetic state, a long-range-ordered state characterized by a checkerboard pattern of alternating spin polarization that breaks spin-rotational symmetry [2, 14]. The excitations of the Néel state are *magnons*—gapless $S = 1$ bosons (also known as spin waves). The four-spin Q-term drives a quantum phase transition to a valence-bond solid (VBS) state [2]. A VBS is a long-range-ordered non-magnetic state formed when sites pair up with their neighbors to form singlets in a well-defined pattern, breaking Z_4 lattice symmetry but not spin-rotational symmetry [2]. An example of a VBS-ordered state can be found in Fig. 4.1, where each blue ellipse represents a singlet pair $(|\uparrow\downarrow\rangle - |\downarrow\uparrow\rangle)/\sqrt{2}$.

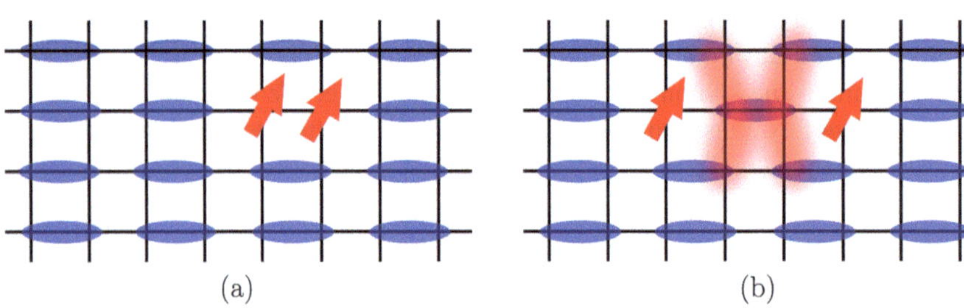

Fig. 4.1 Examples of 2D valence-bond solid (VBS) configurations on an $S = 1/2$ square lattice. Blue ellipses represent singlet pairs, and arrows represent unpaired spins. In (**a**) one singlet bond has been broken to produce a triplet excitation (i.e., a triplon). In (**b**), this triplon has broken into two *spinons*; here the red haze indicates the energy cost of a domain wall between competing VBS orders

This state is sometimes called a "bond order wave" [15], a "spin-Peierls phase," or a "dimerized phase" [16]. This type of state usually occurs in frustrated systems [17, 18] most of which suffer from the sign problem; the J-Q_2 model provides a sign-problem-free way to study the VBS.

The excitations of the VBS are triplons—gapped triplet waves formed by breaking a singlet bond to form a triplet (see Fig. 4.1a). Each triplet is composed of two exotic quasiparticles called *spinons*: spin-1/2 bosons. Within the VBS phase these spinons are confined—pulling them apart generates large numbers of unsatisfied bonds, as in Fig. 4.1b. Thus, like quarks, they never exist as independent particles [19]. At the Néel–VBS transition point, these spinons become *deconfined*, i.e., independently propagating quasiparticles. The actual picture is somewhat more complicated than the cartoon in Fig. 4.1b; each spinon is an unpaired spin at the nexus of four domain walls separating the four degenerate VBS-ordered states [8, 9]. In this sense spinons are vortices of the VBS order parameter [8, 9].

The quantum phase transition from the Néel state to the VBS is of great interest as an example of deconfined quantum criticality (DQC).[2] In the Landau–Ginzburg paradigm phase transitions are described in terms of an order parameter that also describes the ordered phase. Phase transitions between two states that break unrelated symmetries (and thus have different order parameters) are expected to be first order or require fine tuning (i.e., the j_c that destroys the Néel order is exactly the same as the j_c required to produce VBS order by coincidence). Thus, if we discount the possibility of a finely tuned multicritical point, then Landau–Ginzburg theory would predict that either the Néel-VBS transition is first order, or the two phases are separated by an intermediate phase [20]. This transition has been extensively studied in the J-Q model, with substantial evidence that the transition is direct (i.e., no intermediate phase), and so far no sign of any first-order discontinuities [2, 4–7, 21–23]. This violates the prediction of the Landau–Ginzburg theory. The solution to this apparent contradiction is *deconfined quantum criticality*, where the phase transition is governed by a set of fractionalized objects that are confined in both ordered phases (i.e., do not exist as independent objects) [1, 20]. In the case of the J-Q model, the fractionalized excitations are spinons. Spinons appear as independently propagating objects only at the quantum critical point; within the Néel state they are confined within magnons, and within the VBS state they are combined within triplons (gapped magnons). The critical point is therefore described by a theory of spinons, and everywhere else is described by a theory of either magnons or triplons.

4.1.2 *Direct Evidence of Spinons*

It is still a matter of some debate as to whether the Néel-VBS transition is in fact an example of deconfined quantum criticality; there is strong numerical evidence for a

[2] Another example of a proposed non-Landau transition occurs in quantum dimer models [24].

direct continuous phase transition and deconfined quantum criticality [2–7, 21, 23], but it is, of course, difficult to completely rule out the possibility of a weakly first-order transition, and there are some studies showing evidence of a weakly first-order transition [25, 26], although none showing evidence of discontinuous behavior. This transition has already been extensively studied in the J-Q_2 and J-Q_3 models at zero field [2, 4–7, 21–23]. Here we introduce an external Zeeman field, h:

$$H_{JQh} = H_{JQ} - h_z \sum_i S_i^z. \tag{4.3}$$

By adding a field, we force a macroscopic density of magnetic excitations in the ground state. This is a natural continuation of previous work studying the J-Q chain in an external field [27]. We will present evidence that these magnetic excitations are *spinons* rather than conventional *magnons*. Based on field theory arguments, Scammell and Sushkov have predicted [10] that BEC of deconfined spinons should have a unique gapless quadratic mode that in turn produces an anomalous leading-order temperature dependence of specific heat $C_v \propto T$. More recently [11, 13], they have predicted that a gas of deconfined spinons should have a similar gapless quadratic mode and thus also exhibit anomalous specific heat. This anomalous specific heat can *only* be explained by the presence of deconfined spinons, since conventional (nonfractional) magnetic excitations (like magnons) will have no such gapless quadratic mode [10, 11]. Thus, the anomalous specific heat provides a "smoking gun" for the presence of deconfined spinons.

4.1.3 BKT Transition

The Berezinskii–Kosterlitz–Thouless (BKT or sometimes simply KT) transition is an "infinite order" phase transition brought on by the formation of bound pairs of vortices and antivortices.[3] The BKT phase does not have true long-range order or fully broken symmetry. Instead, it has "quasi-long-range order" (QLRO), where correlations decay with a power law. This kind of phase transition was first discovered in the classical XY model (which has $U(1)$ symmetry) [12]. With no external field, the Heisenberg (or J-Q) model has full 3D rotational symmetry $O(3)$. By the Mermin–Wagner Theorem [28], it therefore experiences no finite-temperature phase transitions that break a continuous symmetry, i.e., no long-range spin order. The introduction of the field reduces the symmetry from $O(3)$ to $U(1)$,[4] producing an effective easy-plane AFM and resulting in a finite-temperature BKT

[3] The BKT transition was the subject of the 2016 Nobel Prize in Physics. I thank the Nobel Committee for ensuring that this phase transition was in mind when I was working out what causes the non-monotonic behavior of magnetization and the anomalous specific heat theory to fail at low temperature.

[4] In this case, also equivalent to $O(2)$.

transition to a state with power-law antiferromagnetic correlations in the XY plane. The primary relevance of the BKT transition to this work is that it corresponds to the onset of the spinon BEC, but we will also show that the BKT transition produces non-monotonic behavior in both the uniform and staggered S_z magnetization.

4.1.4 *Outline*

The rest of this chapter will proceed as follows. In Sect. 4.2, we will briefly describe the numerical methods to be used. Section 4.3 contains a description of the phase diagram of the J-Q-h model and the regions characterized in this work. We will then discuss the field-induced BKT transition in Sect. 4.4, where we will estimate $T_{\mathrm{BKT}}(h)$ and discuss the non-monotonic temperature dependence of magnetization. In Sect. 4.5, we will describe the origin of the anomalous specific heat, derive subleading corrections, and discuss results from our quantum Monte Carlo study. Finally, we will discuss conclusions and future work in Sect. 4.6.

4.2 Methods

Here we use the stochastic series expansion (SSE) quantum Monte Carlo (QMC) method [29, 30] with directed loop updates [31] and β doubling [32]. The code used here (also used for Chap. 3) incorporates quantum replica exchange [15, 27, 33], a parallelized multicanonical method where we run many simulations in parallel with different magnetic fields and periodically allow them to stochastically swap fields in a manner that obeys detailed balance. For most of the figures in this chapter, we are interested only a handful of magnetic fields, and the relevant temperatures are high enough that simulations equilibrate well, so we will not use quantum replica exchange. A full description of the Monte Carlo methods employed here can be found in Chap. 5. In Sect. 4.5, we are interested in the temperature dependence of specific heat, but we primarily frame our discussion in terms of the temperature dependence of *energy*. In stochastic series expansion quantum Monte Carlo energy is directly measured whereas specific heat must be calculated either from a discrete derivative of energy (which introduces a discretization error and increases statistical error) or from the fluctuations of the energy, $C_v \propto \langle E^2 \rangle - \langle E \rangle^2$ (which involves a difference of large numbers and is subject to large statistical uncertainties, especially at low temperatures). The measurement of energy is direct and free from any approximation; the only sources of error are statistical error and finite-size effects present in all Monte Carlo results.

4.3 Phase Diagram

In Fig. 4.2 we discuss a zero-temperature phase diagram of the 2D J-Q-h model. This phase diagram is broadly similar to phase diagram for the J-Q-h chain in Fig. 2.2. The h-axis corresponds to the conventional Heisenberg model with an external field. The q-axis is the previously studied zero-field J-Q model [2–7, 21–23]. The point q_c denotes the $T = 0$ quantum phase transition between the Néel and VBS states. This point will also be referred to as j_c where $j \equiv J/Q$ or the deconfined quantum critical (DQC) point ($q \equiv Q/J$). There are several values in the literature for j_c, including recent high-precision results[5]: $j_c = 0.04494(9)$ [6] and $j_c = 0.04468(4)$ [7], and older lower-precision results: $j_c = 0.044$ [4] and $j_c = 0.039(1)$ [3]. The consequence of not being exactly at j_c would be the reappearance of either Néel or VBS order at low temperature (and therefore the end of the quantum critical behavior). Here we will use $j_c = 0.045$. In the presence of a field the precise value of j_c may also change, although we do not study that here. In this chapter we will focus on the deconfined quantum critical (DQC) region in the neighborhood of the point q_c at *finite* temperature; other aspects of this phase diagram relating to the saturation transition are discussed in Chap. 3 (see also [34]).

In order to understand finite-temperature properties we must expand Fig. 4.2 to include a third axis: temperature. For simplicity, we will eliminate the q-axis by setting $q = q_c$ and draw a phase diagram in the T-h plane; we do this in Fig. 4.3. The origin of Fig. 4.3 corresponds to the zero-field, zero-temperature point marked q_c in Fig. 4.2 (the deconfined quantum critical point). The h-axis of Fig. 4.3 (now horizontal) corresponds to a slice of Fig. 4.2 starting from q_c and extending vertically upwards. The T-axis depicts finite temperature (not shown in Fig. 4.2). Along the h axis ($T = 0$), there is a long-range ordered (LRO) phase with spin

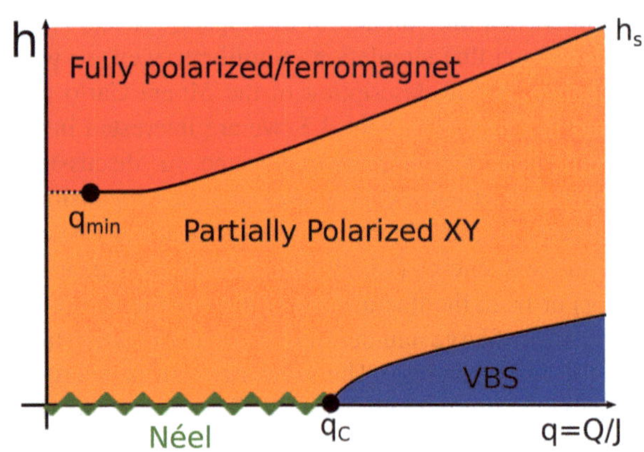

Fig. 4.2 Schematic zero-temperature phase diagram for the J-Q model with external field h. The phases and transitions between them are discussed in the text

[5]Here the number in parenthesis indicated the uncertainty (to one σ) in the digit before it: $1.1(5) \Rightarrow 1.1 \pm 0.5$.

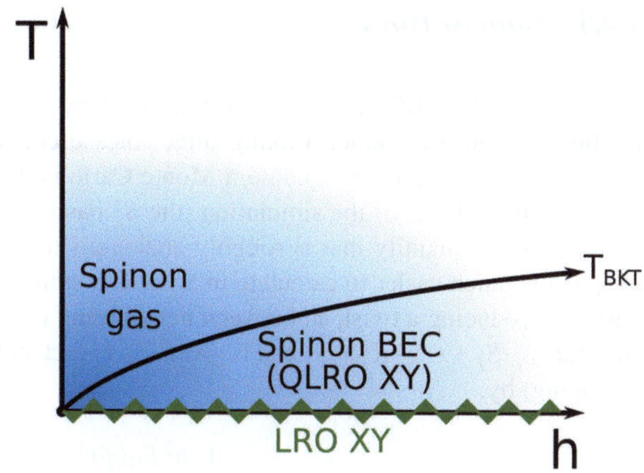

Fig. 4.3 Schematic phase diagram of the J-Q model at criticality ($q = q_c$) as a function of temperature, T, and external field, h. The phases and transitions between them are discussed in the text

correlations in the XY-plane. At finite temperature there is no long-range Néel order since spontaneous breaking of a continuous symmetry is forbidden by the Mermin–Wagner Theorem [28]. Consequently, for $T > 0$, $h = 0$ there is no prevailing long-range order, and at low T there is deconfined quantum criticality. Introducing the magnetic field in the z-direction reduces the full spherical rotational $SO(3)$ symmetry to 2D rotational $U(1)$ symmetry. For $T > 0$, LRO is still prohibited by Mermin–Wagner; instead there is a Berezinskii–Kosterlitz–Thouless-like (BKT) transition to a state with power-law antiferromagnetic spin correlations and finite spin stiffness below $T_{\mathrm{BKT}}(h)$ (this is the spinon quasi-BEC) [12]. For $T > T_{\mathrm{BKT}}$ deconfined spinons are still present, but in a disordered gas-like phase rather than a BEC. T_{BKT} increases with h at low field; at extremely high field, T_{BKT} should eventually go to zero as $h \to h_s$.[6]

4.4 Field-Induced BKT Transition

We will make a rough estimate of $T_{\mathrm{BKT}}(h)$ using the spin stiffness, ρ_s. We will also discuss how the BKT transition causes non-monotonic temperature dependence in both the uniform $\langle m_z \rangle$ and staggered $\langle m_s^2 \rangle$ magnetization in the S^z direction. Since we are only interested in the behavior at the deconfined quantum critical point, we will restrict our study of the BKT transition to the quantum critical point: $j = j_c$. A thorough analysis of the BKT phase boundary as function of h and j and possible coexistence of VBS and Néel order remains a promising subject for a future study.

[6]It is very likely T_{BKT} goes to zero at $h = h_s$ since the system becomes fully magnetized at zero temperature and there are no leftover degrees of freedom to form XY correlations.

4.4.1 Spin Stiffness

The spinon BEC/BKT phase is characterized by power-law decaying spin correlations in the XY plane. Making direct measurements of these correlations in stochastic series expansion quantum Monte Carlo is difficult because they are off-diagonal in the basis of the simulation (the S^z basis). Instead, we will use the spin stiffness, ρ_s, a quantity that is roughly analogous to an elastic modulus of a solid [30] and is much easier to calculate in SSE. The spin stiffness measures the energy cost of introducing a twist, ϕ, between neighboring rows of spins around the z-axis such that $\vec{S}_i \cdot \vec{S}_j \rightarrow \vec{S}_i \cdot R(\phi)\vec{S}_j$ and is given (up to factors having to do with rotational averaging) by:

$$\rho_s = \frac{1}{L^2} \frac{\partial^2 E_0(\phi)}{\partial \phi^2}\bigg|_{\phi=0} \tag{4.4}$$

where $E_0(\phi)$ is the ground state energy with twist ϕ [32, 35]. The spin stiffness is not a direct measure of spin correlations, but it *is* a way to differentiate slowly decaying correlations (i.e., LRO or QLRO) from rapidly decaying correlations (i.e., exponential). In states with exponentially decaying (short-range) spin correlations, such as the VBS, the spin stiffness vanishes in the thermodynamic limit, but the spin stiffness is finite in states with LRO or QLRO spin order such as the Néel or BKT states.[7]

In the SSE formulation, the spin stiffness can be calculated from fluctuations of the *winding number*. We can map the spin configuration onto hardcore bosons, where spin-down sites are empty and spin-up sites are occupied by spin-one bosons. SSE is a path-integral formulation with periodic boundary conditions in the imaginary time direction, so the final state must be identical to the original state. If we trace the world-lines of these bosons, they can all connect back to themselves, or they can wrap around the periodic boundaries (in space) to a different boson (since they are hardcore bosons, the world-lines cannot cross). For example, in a chain the world-line for the first boson could connect to the second boson in the final state, the second to the third, and so on until the last boson wraps around the spatial boundary to connect to the first. For an example of SSE configurations with zero and nonzero winding numbers see Fig. 50 of [30]. The winding number in the $+\hat{x}$ direction is the difference between the number of rightward and the number of leftward moves of these bosons:

$$W_x = \frac{N_x^R - N_x^L}{L_x} \tag{4.5}$$

[7] For a discussion of the spin stiffness in the 2D Heisenberg antiferromagnet as a function of field, see [36].

where N_x^R is the total number of x-direction operators of the form $\frac{-+}{+-}$ and N_x^L is the number of x-direction operators of the form $\frac{+-}{-+}$ [32]. W_x then takes on integer values $0, \pm 1, \pm 2 \ldots$. If $W_x \neq 0$, then there is a net "current" of bosons around the spatial boundary. The spin stiffness is then extracted from the fluctuations of this current [32]:

$$\rho_s = \frac{\langle W_x^2 \rangle + \langle W_y^2 \rangle}{2\beta} \tag{4.6}$$

in two dimensions.[8] The winding number is easily calculated from the operator string and spin configuration in stochastic series expansion quantum Monte Carlo.

In Fig. 4.4, we examine the finite-size scaling of $\rho_s(T)$ with $j = j_c$ for a few representative values of magnetic field; the axes of all four panels are identical to

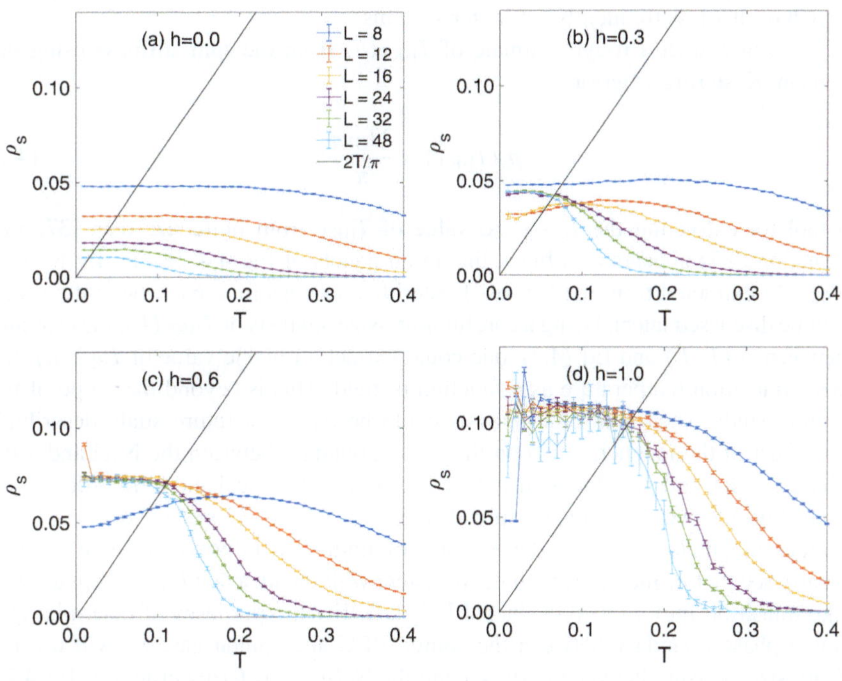

Fig. 4.4 Finite-size scaling of spin stiffness, ρ_s, of the J-Q-h model a few representative magnetic fields with $j = j_c = 0.045$. The color lines are QMC results without quantum replica exchange. The black lines represent the Nelson–Kosterlitz criterion, Eq. (4.7). In (**a**), the $h = 0$, ρ_s is finite, but approaches zero as $T \to 0$. With nonzero magnetic field in (**b**) $h = 0.3$, (**c**) $h = 0.6$, and (**d**) $h = 1.0$, ρ_s is finite for low T and approaches a finite value as $L \to \infty$

[8] Again, there may be additional factors having to do with rotational averaging to match the strictest definition of ρ_s, but this is the formula used in our calculations.

allow for direct comparison.[9] The zero-field case is depicted in Fig. 4.4a; here the stiffness clearly approaches zero as $L \to \infty$. For intermediate fields, Fig. 4.4b–d, ρ_s converges toward a finite value for large L, but exhibits some non-monotonic, cross-over-like, behavior as a function of L, which is a consequence of the proximity to other phases (especially the VBS). In the thermodynamic limit, phase boundaries are sharp, but at finite size the phase boundary is fuzzy and the system can exhibit behavior corresponding to multiple phases. This can be resolved through application finite-size scaling analysis.

At high field, $h = 1.0$ in Fig. 4.4d, ρ_s is larger and quickly converges to a finite value as $T \to 0$ and $L \to \infty$. At higher fields, larger sizes, and lower temperatures, the uncertainty in ρ_s becomes large. Even for $h = 1.0$ in Fig. 4.4d we see some signs that it is failing to equilibrate (where the fluctuations appear substantially larger than the error bars). This has a simple explanation in terms of simulation dynamics: the spin stiffness is extracted from the winding number, which is a global quantity; updating a global quantity requires a loop to wrap all the way around the system, which occurs less frequently for larger systems.

We can extract a rough estimate of $T_{\mathrm{BKT}}(h)$ from the spin stiffness using the Nelson–Kosterlitz criterion,

$$\rho_s(T_{\mathrm{BKT}}) = \frac{2T_{\mathrm{BKT}}}{\pi}, \tag{4.7}$$

a tool for extracting the $L \to \infty$ value of T_{BKT} from finite-size data [37, 38]. Equation (4.7) appears as a black line in all panels of Fig. 4.4. In Fig. 4.9 we plot $T_{\mathrm{BKT}}(h)$ extracted from the Nelson–Kosterlitz crossings for a 64×64 system (this will be discussed later). Using a careful finite-size analysis of $T_{\mathrm{BKT}}(L)$, the crossing between $\rho_s(T, L)$ and Eq. (4.7), one could extract a reliable value of $T_{\mathrm{BKT}}(h)$, the BKT transition temperature as a function of field. This is beyond the scope of the current study; such a careful analysis could be part of a future study describing the affect of the magnetic field on the phase boundary between the Néel and VBS phases, $j_c(h)$, and possible field-induced coexistence of Néel and VBS order.

In Fig. 4.5, we consider the spin stiffness for a 32×32 system at $j = j_c$ for various magnetic fields. At the expense of finite-size information, here we can clearly see that ρ_s increases with h, consistent with the idea that h is driving a BKT-like transition into a QLRO/spinon BEC state. For the purposes of determining a rough phase boundary between the spinon BEC and spinon gas, we will use the finite-size crossing between $\rho_s(h, T)$ and the Nelson–Kosterlitz criterion (Eq. 4.7), as a rough estimate of $T_{\mathrm{BKT}}(h)$. We plot $T_{\mathrm{BKT}}(h)$ extracted in this manner for a 64×64 system in Fig. 4.9.

[9] For a discussion of the magnetic field effect on the spin stiffness in the Heisenberg model, see [36].

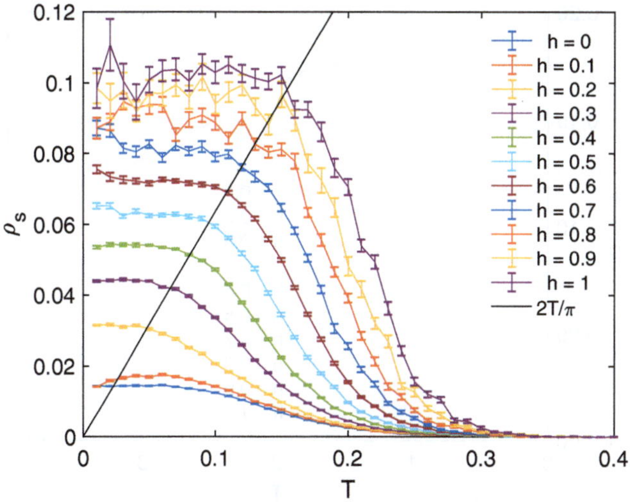

Fig. 4.5 Spin stiffness as a function of temperature in the 2D J-Q model with $j = j_c = 0.045$ on a 32×32 square lattice with various magnetic fields using QMC without quantum replica exchange. The black line is the Nelson–Kosterlitz criterion, Eq. (4.7)

4.4.2 Non-monotonic $m(T)$ Dependence

In Fig. 4.6 we plot $\langle m \rangle / h$ (the magnetization divided by the field) as a function of temperature for a 64×64 system for $j = j_c$, where m is the uniform z-direction magnetization density defined:

$$m \equiv \frac{2}{L^2} \sum S_i^z. \tag{4.8}$$

At high temperature, dividing by h produces the expected collapse onto a single curve, with $m/h \to 0$ as $T \to \infty$. However, around $T \approx 2$, the magnetization curves separate, indicating that the magnetization is no longer linear in the field. For lower temperatures, magnetization decreases before experiencing a local minimum in temperature and then approaching its zero-temperature value. This non-monotonic temperature dependence of magnetization is a signature of the BKT transition. Although it has been documented in the literature [39–44], this fact is not widely understood in the quantum magnetism community so it is worth discussing further here.

We examine the low-temperature regime of Fig. 4.6 in Fig. 4.7. At low temperature, the magnetization has a super-linear dependence on the field, i.e., m grows faster than $m \propto h$. We also observe that $m(T)$ is non-monotonic, with a local minimum at a finite temperature, $T_{\min}(h)$. The minimum in $m(T)$ for each value of h in Fig. 4.7 is marked with black X's (except for $h = 0.1$, where the minimum could not be resolved). A similar effect has been noted in the pure 2D Heisenberg

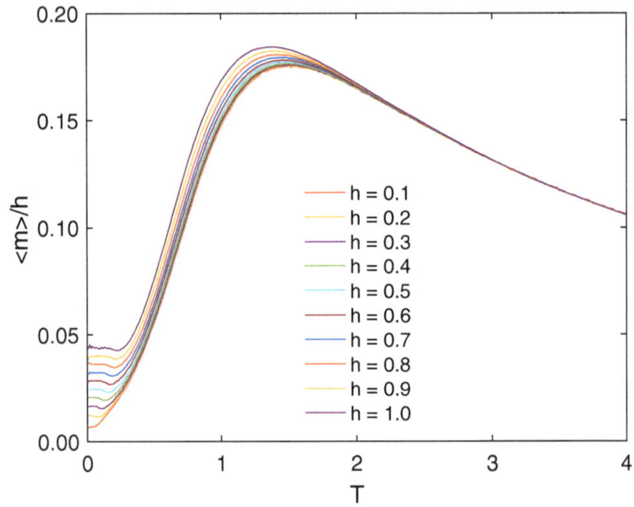

Fig. 4.6 Magnetization divided by field, $\langle m \rangle / h$, as a function of temperature for the 2D J-Q model on a 64×64 lattice with $j = j_c = 0.045$ using QMC without quantum replica exchange. Markers and error bars omitted for clarity. If color is not available, the field increases from the bottom curve ($h = 0.1$) to the top curve ($h = 1$). An enlarged view of the low-temperature region is shown in Fig. 4.7

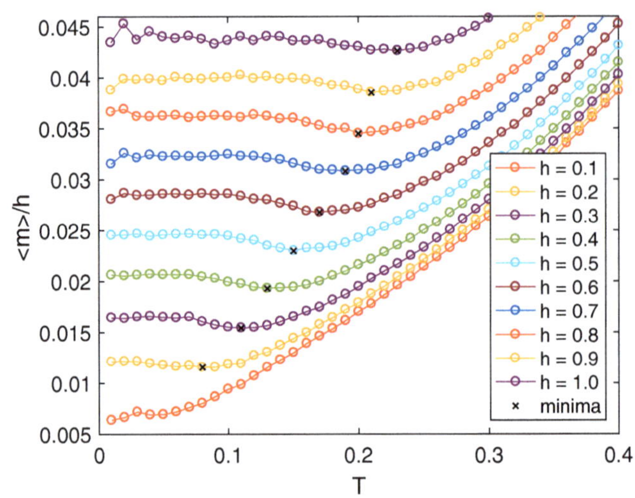

Fig. 4.7 Magnetization divided by field, $\langle m \rangle / h$, as a function of temperature for the 2D J-Q model on a 64×64 square lattice with $j = j_c = 0.045$ using QMC without quantum replica exchange. m is the uniform z-direction magnetization density defined in Eq. (4.8). $\langle m \rangle / h$ exhibits non-monotonic temperature dependence with finite-temperature minima marked by black X's. Error bars were in all cases smaller than the markers. The high temperature behavior is depicted in Fig. 4.6

AFM in both the classical [39, 40] and quantum cases [41–43],[10] and has also been reported in experiments [44, 45]. This same effect also occurs in the absence of a field in both the classical [46] and quantum [47] AFM 2D Heisenberg models with intrinsic anisotropy, which also have a BKT transition.

A rough explanation of the non-monotonic behavior is as follows: at extremely high temperatures, the magnetization vanishes because the field (and the rest of the Hamiltonian) is washed out by thermal fluctuations. As the temperature decreases, the bias from the field "turns on" and the magnetization increases. As the temperature approaches T_{BKT}, there is an onset of antiferromagnetic power-law correlations. These correlations are primarily in the XY plane, but they compete with the field, thus reducing the uniform magnetization. For $T < T_{\mathrm{min}}$, further reducing the temperature suppresses thermal fluctuations, this allows the uniform magnetization to recover slightly as $T \to 0$. At exactly zero temperature, the ground state is a canted AFM where spins are tilted by some angle θ from the z axis and there are long-range spin correlations in the XY plane.

As similar story occurs in the staggered S^z magnetization defined:

$$m_s = \frac{2}{L^2} \sum_{x,y=1}^{L} (-1)^{x+y} S^z(x,y). \tag{4.9}$$

The field does not break the staggered spin inversion symmetry $m_s \iff -m_s$, so $\langle m_s \rangle$ always vanishes. Instead we plot the squared staggered magnetization, $\langle m_s^2 \rangle$, in Fig. 4.8. $\langle m_s^2 \rangle$ is a measure of the antiferromagnetic spin correlations in the S^z direction. At high temperature, it naturally goes to zero. As the temperature is lowered, it has a *maximum* at approximately the same temperature as the *minimum* in the uniform magnetization, and then decreases as $T \to 0$. $\langle m_s^2 \rangle$ vanishes at zero temperature. The field dependence of the (temperature) maximum of $\langle m_s^2 \rangle$ is the same at the field dependence of T_{BKT} and T_{min} because the physics is the same: the competition between the field and the antiferromagnetic BKT correlations.

4.4.3 Estimation of T_{BKT}

In Fig. 4.9, we combine $T_{\mathrm{BKT}}(h)$ (see Fig. 4.5) and $T_{\mathrm{min}}(h)$ extracted from $m(T)$ (see Fig. 4.7). The BKT transition temperature is extracted from the spin stiffness of a 64×64 J-Q system using the Nelson–Kosterlitz criterion, Eq. (4.7). We know that T_{BKT} must vanish at zero field, where full rotational symmetry is restored and the BKT transition is therefore impossible, and it will most likely return to zero at saturation, where at zero temperature there are no degrees of freedom left to order. Indeed, in Fig. 4.9 both T_{BKT} and T_{min} approach zero as $h \to 0$. In the case of T_{min}, it was not possible to reliably extract a finite-T minimum in $m(T)$ for $h \leq 0.1$. For T_{BKT}, there is a small nonzero value of $T_{\mathrm{BKT}}(h = 0)$. This is a result of finite-size effects, and we expect that $T_{\mathrm{BKT}}(h = 0)$ will go to zero in the thermodynamic limit.

[10] See Fig. 4 of [41].

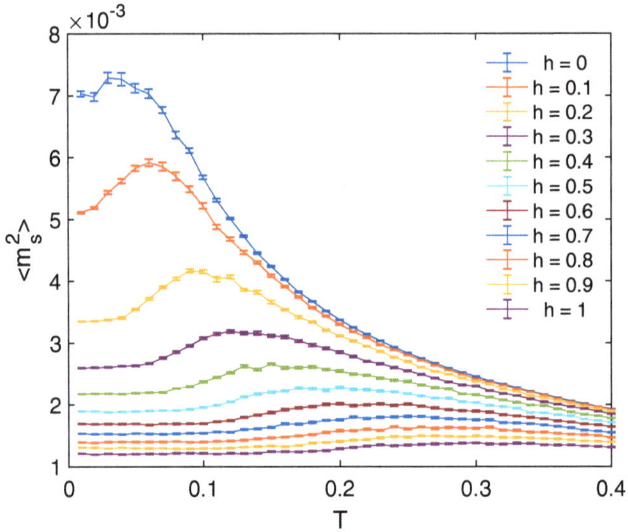

Fig. 4.8 Squared staggered magnetization, $\langle m_s^2 \rangle$, for a 64 × 64 J-Q system tuned to criticality $j = j_c = 0.045$ for various magnetic fields. Results from quantum Monte Carlo without quantum replica exchange

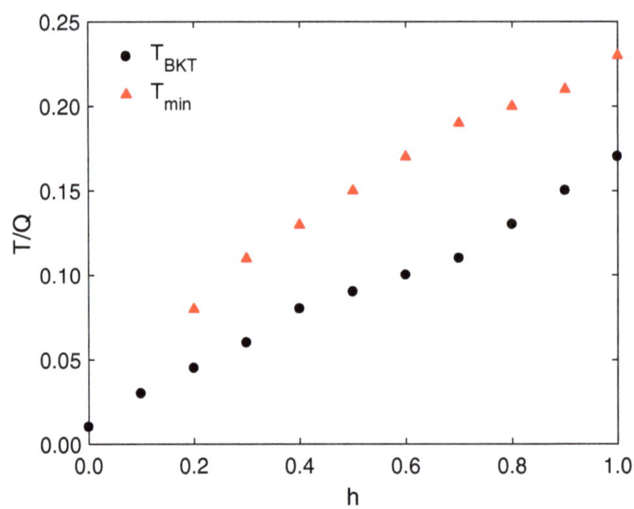

Fig. 4.9 Estimation of T_{BKT} (circles) and $T_{\min}$ (triangles, the finite-T minimum in $m(T)$) based on a 64 × 64 J-Q system tuned to $j = j_c = 0.045$ for various magnetic fields. Results from quantum Monte Carlo without quantum replica exchange

We also observe that $T_{\min} > T_{\mathrm{BKT}}$. As the quantities in Fig. 4.9 are only rough estimates; error bars are not included. Here we are only interested in (1) providing a rough phase boundary for the spinon BEC and (2) checking that these quantities (T_{BKT} and $T_{\min}$) make physical sense.

4.5 Anomalous Specific Heat

It has already been shown [5] that for the zero-field case, the leading temperature dependence of the specific heat at the deconfined quantum critical (DQC) point is quadratic in temperature. Starting from the DQC point in the J-Q model, we will add a magnetic field, inducing a finite density of magnetic excitations. These magnetic excitations will form a Bose–Einstein condensate (BEC).[11] The specific heat of a BEC of magnons (conventional spin-1 quasiparticles) would vary *quadratically* with temperature (to leading order) [10]. A BEC of spinons (unconventional spin-1/2 quasiparticles), however, is expected [10] to have a unique gapless quadratic mode, which in turn produces an anomalous *linear* contribution to the specific heat:

$$C_v = \frac{\zeta(2)}{\pi c^2} hT \tag{4.10}$$

in the limit $T \ll h \ll Q$ [10, 11], where $\zeta(2) = \frac{\pi^2}{6}$ is Riemann's Zeta function and c is the excitation velocity, which should be independent of both field and temperature. For c we use the value from [6] where $\frac{c}{J+Q} = 2.31(5)$, which for $J = 0.045$, $Q = 1$ yields $c = 2.42$.[12]

Although Eq. (4.10) provides the leading-order contribution to the specific heat, we also expect to see large subleading contributions. With this in mind, we will now describe the origin of the anomalous temperature dependence and then derive the contributions to the energy from both gapless spinon modes in full detail. This linear temperature dependence in Eq. (4.10) arises from just one of the four modes of the $S = 1/2$ BEC, given by [10]:

$$\omega_1 = \sqrt{h^2 + c^2k^2} - h \tag{4.11a}$$

$$\omega_2 = \sqrt{3h^2 - m^2 + c^2k^2 - \sqrt{(3h^2 - m^2)^2 + 4h^2c^2k^2}} \tag{4.11b}$$

$$\omega_3 = \sqrt{h^2 + c^2k^2} + h \tag{4.11c}$$

$$\omega_4 = \sqrt{3h^2 - m^2 + c^2k^2 + \sqrt{(3h^2 - m^2)^2 + 4h^2c^2k^2}} \tag{4.11d}$$

Note here that Scammell and Sushkov [10] have ignored the possible contribution from a hypothesized dynamical gauge field based on the fact that it was not necessary to explain the thermodynamics of the zero-field J-Q model at j_c [5].

---

[11] This BEC may not meet the strictest definition of a BEC (true long-range off-diagonal order), since such ordering is forbidden in 2D, but there will be power-law correlations. This BEC will therefore still be a "stiff" phase, so it is a BEC for our purposes.

[12] Other studies have claimed different values for the spinon velocity such as $c = 2.4(3)$ (with $Q = 1$) [48] and $c = 2.55$ (with $Q = 1$) [5].

Modes ω_1 and ω_2 are both gapless (Goldstone) modes. If we expand ω_1 around $k \ll h$

$$\begin{aligned}\omega_1 &= \sqrt{h^2 + c^2k^2} - h = -h + h\sqrt{1 + \frac{c^2k^2}{h^2}} \\ &= -h + h\left(1 + \frac{c^2k^2}{2h^2} - \frac{c^4k^4}{8h^4} + \cdots\right) \\ \omega_1 &\approx \frac{c^2k^2}{2h}. \end{aligned} \tag{4.12}$$

We see that it is gapless and quadratic in k. If we then expand ω_2 for small k:

$$\begin{aligned}\omega_2 &= \sqrt{3h^2 - m^2 + c^2k^2 - \sqrt{(3h^2 - m^2)^2 + 4h^2c^2k^2}} \\ &= \sqrt{b + c^2k^2 - \sqrt{b^2 + ac^2k^2}} \\ &= \sqrt{b + c^2k^2 - b\left(1 + \frac{ac^2k^2}{2b^2} + \cdots\right)} \\ &\approx \sqrt{b + c^2k^2 - b - \frac{ac^2k^2}{2b}} = ck\sqrt{1 - \frac{a}{2b}} \\ \omega_2 &\approx ck\sqrt{\frac{h^2 - m^2}{3h^2 - m^2}} = vk \end{aligned} \tag{4.13}$$

We find that ω_2 is a gapless linear mode with modified spinon velocity v defined:

$$v \equiv c\sqrt{\frac{h^2 - m^2}{3h^2 - m^2}} \tag{4.14}$$

Modes ω_3 and ω_4 are both gapped, with gaps

$$\Delta_3 = 2h \tag{4.15a}$$

$$\Delta_4 = \sqrt{2(3h^2 - m^2)}. \tag{4.15b}$$

so they will (to leading order) be unoccupied at low temperature $T \ll \Delta_3, \Delta_4$. In order to exclude these subleading modes, h must be large. Next we will show that the quadratic dispersion of ω_1 produces the anomalous linear contribution to the specific heat.

4.5.1 Full Contributions from the Gapless Modes

To find the leading contribution to the specific heat, Scammell and Sushkov considered the low-temperature limit $T \ll h$. Subleading contributions to the specific heat of the spinon BEC have large prefactors. We will for now ignore the contributions from the gapped modes ω_3 and ω_4 and calculate the full (exact) contributions from the gapless modes (ω_1 and ω_2). The full (exact) temperature dependence of the energy (and therefore specific heat) of the spinon BEC can be derived by integrating the Bose–Einstein occupation function,

$$E_i = V \int_0^\infty \frac{\hbar\omega_i g(\omega_i) d\omega_i}{e^{\beta\hbar\omega_i} - 1}, \tag{4.16}$$

for each mode [49].

Let us begin with the gapless quadratic mode ω_1 (Eq. 4.11a). Before we can integrate Eq. (4.16), we must derive the density of states for ω_1: $g(\omega_1)$. First we find an expression for Ω—the total number of states with energy $\omega \leq \omega_0$ by integrating over all $\vec{k}$ such that $\omega(\vec{k}) \leq \omega_0$:

$$\Omega(\omega_0) = \Omega(k_0) = \int_0^{\omega(k)\leq\omega_0} \frac{d^d k}{(2\pi)^d}$$

$$\Omega(k_0) = \frac{S_d}{(2\pi)^d} \int_0^{k_0} k^{d-1} dk = \frac{S_d}{(2\pi)^d} \frac{k_0^d}{d} \tag{4.17}$$

$$\Omega(k_0) = \frac{k_0^2}{4\pi} \tag{4.18}$$

Then write Ω in terms of ω_1:

$$\omega_1 = \sqrt{h^2 + c^2k^2} - h$$

$$c^2k^2 = \omega_1^2 + 2h\omega_1 \tag{4.19}$$

using the exact form of the dispersion from Eq. (4.11a). Plug this into Eq. (4.18) and take the derivative to find the density of states:

$$g(\omega_1) = \frac{d\Omega(\omega_1)}{d\omega_1} = \frac{1}{2\pi c^2}(\omega_1 + h) \tag{4.20}$$

We can find the energy in the ω_1 mode by integrating Eq. (4.16), setting $\hbar = k_B = 1$:

$$\frac{E_1}{V} = \frac{1}{2\pi c^2} \int_0^\infty \frac{\omega_1(\omega_1 + h) d\omega_1}{e^{\beta\omega_1} - 1}$$

and making the substitution $x \equiv \omega_1/T$:

$$\begin{aligned}
\frac{E_1}{V} &= \frac{1}{2\pi c^2}\left[T^3 \int_0^\infty \frac{x^2 dx}{e^x - 1} + hT^2 \int_0^\infty \frac{x dx}{e^x - 1}\right] \\
\frac{E_1}{V} &= \frac{1}{2\pi c^2}\left[\Gamma(3)\zeta(3)T^3 + \Gamma(2)\zeta(2)hT^2\right] \\
\frac{E_1}{V} &= \frac{\pi}{12c^2}T^2 h + \frac{\zeta(3)}{\pi c^2}T^3
\end{aligned} \tag{4.21}$$

where $\Gamma(s)$ is the usual gamma function[13] and $\zeta(s)$ is the Reimann Zeta function[14] defined:

$$\zeta(s) \equiv \frac{1}{\Gamma(s)} \int_0^\infty \frac{x^{s-1} dx}{e^x - 1}. \tag{4.22}$$

Finally, to find the specific heat contribution from ω_1, we take the derivative:

$$C_1 = \frac{\partial}{\partial T}\frac{E_1}{V} = \frac{\pi}{6c^2}hT + \frac{3\zeta(3)}{\pi c^2}T^2 \tag{4.23}$$

Here we find the expected linear contribution to specific heat as in [10], but with a subleading term proportional to T^2. The prefactor for the linear term is actually quite small ≈ 0.1 for $h = 1$, but the prefactor for the quadratic term is twice as large ≈ 0.2. This is not even the full subleading contribution, for that we will need the contribution from ω_2.

To find the contribution for ω_2 we will follow the same procedure as before, starting with the Bose integral from Eq. (4.16). To get the density of states, $g(\omega_2)$, we must solve for $k(\omega_2)$ using Eq. (4.11b):

$$\begin{aligned}
\omega_2 &= \sqrt{3h^2 - m^2 + c^2k^2 - \sqrt{(3h^2 - m^2)^2 + 4h^2c^2k^2}} \\
\omega_2^2 &= 3h^2 - m^2 + c^2k^2 - \sqrt{(3h^2 - m^2)^2 + 4h^2c^2k^2} \\
\sqrt{(3h^2 - m^2)^2 + 4h^2c^2k^2} &= 3h^2 - m^2 + c^2k^2 - \omega_2^2
\end{aligned}$$

[13] $\Gamma(3) = 2, \Gamma(2) = 1$.

[14] $\zeta(2) = \pi^2/6, \zeta(3) \approx 1.20206$.

$$(3h^2 - m^2)^2 + 4h^2c^2k^2 = \left[(3h^2 - m^2) + c^2k^2 - \omega_2^2\right]^2$$
$$0 = c^4k^4 + c^2k^2\left[2h^2 - 2m^2 - 2\omega_2^2\right] + \left[\omega_2^4 - 2\omega_2^2(3h^2 - m^2)\right]$$

Now we can plug into the quadratic equation to arrive at:

$$c^2k^2 = m^2 + \omega_2^2 - h^2 \pm \sqrt{h^4 + m^4 - 2h^2m^2 + 4\omega_2^2h^2} \tag{4.24}$$

We can set the spinon mass, m, to zero, because it vanishes at the deconfined quantum critical point [11], which is the only case that we are interested in.

$$c^2k^2 = \omega_2^2 - h^2 \pm h\sqrt{h^2 + 4\omega_2^2} \tag{4.25}$$

Now we choose the first (+) solution because the (−) solution produces negative values for k^2 for small ω_2. Plug this solution into Eq. (4.18) and then Eq. (4.20) to get the density of states for mode ω_2

$$g(\omega_2) = \frac{1}{2\pi c^2}\left(\omega_2 + \frac{2h\omega_2}{\sqrt{h^2 + 4\omega_2^2}}\right) \tag{4.26}$$

Next we plug the result into Eq. (4.16), set $\hbar = k_b = 1$, and substitute $x \equiv \omega_2/T$

$$\frac{E_2}{V} = \frac{T^3}{2\pi c^2}\int_0^\infty \left(x + \frac{2xh}{\sqrt{h^2 + 4T^2x^2}}\right)\frac{xdx}{e^x - 1}$$
$$\frac{E_2}{V} = \frac{T^3}{2\pi c^2}\left[\int_0^\infty \frac{x^2dx}{e^x - 1} + 2\int_0^\infty \frac{1}{\sqrt{1 + 4T^2x^2/h^2}}\frac{x^2dx}{e^x - 1}\right]$$

Here we cannot write everything in terms of standard Reimann Zeta functions, so we instead write

$$\frac{E_2}{V} = \frac{\zeta(3) + G(y)}{\pi c^2}T^3 \tag{4.27}$$

where we define $y \equiv h/T$ and define the integral (no longer independent of h or T)

$$G(y) \equiv \int_0^\infty \frac{1}{\sqrt{1 + 4\frac{x^2}{y^2}}}\frac{x^2dx}{e^x - 1}. \tag{4.28}$$

As before we can determine the ω_2 contribution to the energy, but we now have the additional complication that the prefactor to the T^3 term is not independent of T. Here we are interested primarily in estimating the prefactor of the T^3 dependence arising from ω_2. If we consider the limit $h \gg T$, then $G(y) \to \Gamma(3)\zeta(3)$, and the relation becomes cubic in temperature:

$$\frac{E_2}{V} \approx \frac{3\zeta(3)}{\pi c^2} T^3. \tag{4.29}$$

From the energy we can then derive the specific heat contribution from mode ω_2:

$$C_2 \approx \frac{9\zeta(3)}{\pi c^2} T^2 \tag{4.30}$$

and see that it has a conventional T^2 leading-order temperature dependence, with a prefactor of ≈ 0.6, again, much larger than the $C_v \propto T$ prefactor of ≈ 0.1 from Eq. (4.23). Although the $C \propto T^3$ terms are subleading, the will still be large when $T = \mathcal{O}(0.1)$.

Hereafter we will focus our analysis on the temperature dependence of the energy rather than the specific heat because measurements of energy are more reliable in our QMC program.[15] We can write the total energy contributions for the gapless modes ω_1 and ω_2:

$$\frac{E_{12}}{V} = \frac{\pi}{12c^2} h T^2 + \left[\frac{2\zeta(3) + G(y)}{\pi c^2} \right] T^3 \tag{4.31}$$

Here the contribution from ω_2 has compounded the earlier problem that the prefactor for the anomalous leading temperature dependence is smaller than the prefactor for the conventional cubic temperature dependence. Note that we have made no approximations between our starting point (the dispersions of the modes) and this expression. This expression is therefore an *exact* description of the energy in these modes (ω_1 and ω_2). It will not, however, exactly describe the energy of the entire spin system, which will naturally include contributions from other degrees of freedom not considered here, including uncondensed spinons. We have also neglected contributions from the gapped modes, ω_3 and ω_4, but in practice these gaps are relatively small (i.e., it will not the case that $T \ll \Delta$). Therefore we expect that modes ω_3 and ω_4 *will* make some non-negligible contribution.

[15]In the SSE, the energy can be extracted directly from the configuration using $\langle E \rangle = -\langle n \rangle / \beta$, where n is the total number of nonidentity operators in the operators string. The specific heat can be extracted from the fluctuations of the energy using $\frac{1}{kT^2}\left(\langle E^2 \rangle - \langle E \rangle^2\right)$ or by taking discretized derivatives, but both of these procedures introduce additional error (statistical in the former case and discretization in the latter). See Sect. 5.4.3 for a more detailed discussion.

4.5.2 QMC Results

We will now test the prediction of Scammell and Sushkov [10] using our quantum Monte Carlo program. We will compare the leading-order temperature dependence of the energy from QMC to the form predicted by theory in Eq. (4.31). In terms of energy, the anomalous temperature dependence appears as a quadratic term ($E \propto T^2$), and the subleading (trivial) contributions as cubic ($E \propto T^3$) and higher-order terms.

In Fig. 4.10, we plot the $E(h, T)$, the energy of a 64×64 J-Q system at the deconfined quantum critical point (here we use $j = j_c = 0.045$ [6, 7]). Here the magnetic field clearly reduces the energy and appears to affect the temperature dependence, but the anomalous contribution cannot be clearly identified below T_{BKT}. To address this issue, we consider $\Delta E(h, T)$, where we subtract the zero-field energy from the finite-field energy to isolate the contribution from the magnetic excitations:

$$\Delta E(h, T) \equiv E(h, T) - E(h = 0, T) + f(h) \tag{4.32}$$

where $f(h)$ is some function of h (but not T) that ensures that the lines do not overlap in Fig. 4.11. We plot $\Delta E(h, T)$ against T^2 in Fig. 4.11 (using the same data as in Fig. 4.10). For clarity, we have omitted error bars (which were in all cases smaller than the markers), and plot only high magnetic fields. For each h the points T_{BKT}, $1.5T_{\text{BKT}}$, and $2T_{\text{BKT}}$ are noted with large black markers. In

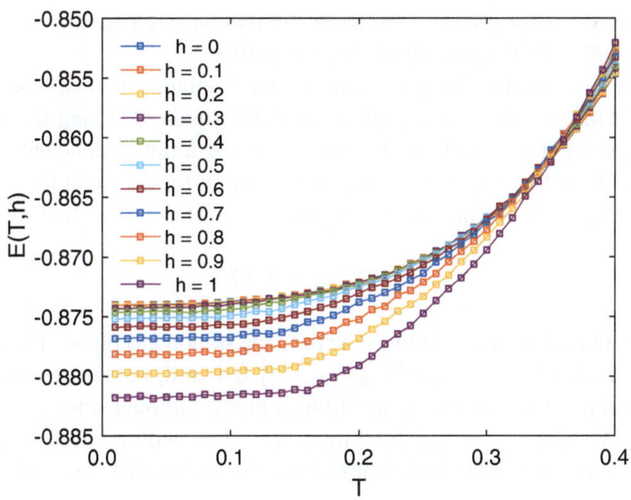

Fig. 4.10 The finite-temperature energy $\langle E \rangle$ for a 64×64 J-Q system tuned to criticality $j = j_c$ for various magnetic fields. Results from quantum Monte Carlo without quantum replica exchange

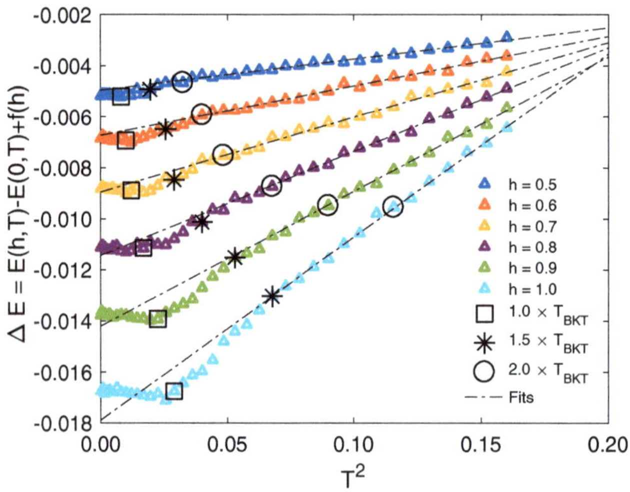

Fig. 4.11 Temperature dependence of energy with the zero-field energy subtracted off to remove subleading corrections and a constant, $f(h)$, added to ensure that lines do not overlap as given in Eq. (4.32). System: 64 × 64 J-Q model with $j = j_c = 0.045$. Large black markers indicate $T_{\rm BKT}$, $1.5T_{\rm BKT}$, and $2T_{\rm BKT}$ for each value of h. Dashed lines represent a fit to the form $\Delta E(h, T) = a + bT^2$ with a lower cutoff of $T_{\rm cut} = 2T_{\rm BKT}$

Fig. 4.11 we can still see no sign of the anomalous temperature dependence below $T_{\rm BKT}$. Even incorporating subleading corrections derived in the previous section (Eq. 4.31) and fitting to polynomials of various orders, we were unable to produce a fit to the behavior of the system that included the expected anomalous temperature dependence ($E \propto T^2$) as a result of the spinon BEC.

Let us now consider the gas of deconfined spinons that appears for $T > T_{\rm BKT}(h)$.[16] This gas also has a gapless quadratic mode [11], and therefore would also be expected to exhibit an anomalous temperature dependence $E \propto T^2$, although we do not have a detailed theory prediction for the exact form of $E(T, h)$. Looking at Fig. 4.11, we indeed see this behavior. The dashed lines are fits to the form

$$\Delta E(T) = a + bT^2 \tag{4.33}$$

with a lower bound of $2T_{\rm BKT}$ to limit the fit to the spinon gas phase. These fits match the data within the fitting range $2T_{\rm BKT} \leq T \leq 0.4$ well, but are also a good match for lower temperatures (outside the fitting range), although they cease to match near $T_{\rm BKT}$. We also performed this fitting procedure with a cutoff temperature of $1.5T_{\rm BKT}$, which produced fit lines that were nearly impossible to visually distinguish

[16]Here I must thank Harley Scammell and Oleg Sushkov for helping to correctly identify the region $T > T_{\rm BKT}(h)$ as corresponding to a gas of spinons and the low-temperature region as corresponding to the BEC of spinons.

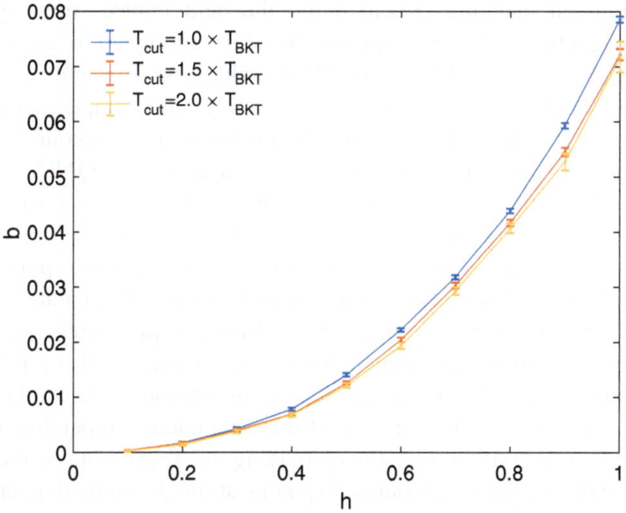

Fig. 4.12 Slope of fit to $\Delta E(h, T) = a + bT^2$ for a lower cutoff of $T_{\text{cut}} = T_{\text{BKT}}$, $T_{\text{cut}} = 1.5T_{\text{BKT}}$ and $T_{\text{cut}} = 2T_{\text{BKT}}$, using data from Fig. 4.11

from those plotted in Fig. 4.11 (we therefore do not present those fits here, but the resulting fitting parameters appear in Fig. 4.12).

In Fig. 4.12, we plot, as a function of h, the coefficient b from fitting $\Delta E(h, T)$ to the form Eq. (4.33) with an upper cutoff of $T = 0.4$ and lower cutoffs of $T_{\text{cut}} = T_{\text{BKT}}$, $T_{\text{cut}} = 1.5T_{\text{BKT}}$, and $T_{\text{cut}} = 2T_{\text{BKT}}$. The fitting parameters $b(h)$ for lower cutoffs of $1.5T_{\text{BKT}}$ and $2T_{\text{BKT}}$ agree within error bars. For $T_{\text{cut}} = T_{\text{BKT}}$, $b(h)$ is a bit further from the other two, but is a reasonable qualitative match. The fact that the fits are rather insensitive to the cutoff is a good indication that the form Eq. (4.33) accurately describes the behavior of the system and the $E \propto T^2$ contribution is real. Here we are not looking for a dominant $E \propto T^2$ contribution, but *any* such contribution, since spinons are the only mechanism that can produce an $E \propto T^2$ contribution. All other scenarios, such as magnons or the zero-field DQC point produce a leading temperature dependence of $E \propto T^3$. The $E \propto T^2$ contribution for $T > T_{\text{BKT}}(h)$ is therefore direct evidence for the presence of a gas of deconfined spinons.

4.6 Conclusions

Here we have studied the magnetic field effect in a 2D quantum antiferromagnet—the J-Q model—focusing on the region around the deconfined quantum critical point. The introduction of the field sets the stage for a finite-temperature BKT phase characterized by power-law spin correlations in the XY plane. The BKT transition causes non-monotonic temperature dependence in the magnetization.

At the deconfined quantum critical point, the field induces finite ground state density of magnetic excitations—spinons. We have attempted to test a field theory prediction by Scammell and Sushkov [10] that the Bose–Einstein condensation of these spinons should produce an anomalous leading-order temperature dependence of energy $E \propto T^2$. We did not verify this prediction; even accounting additional subleading terms, we were unable to produce a fit to our QMC data that was consistent with the theory prediction for a spinon BEC. We now suspect that the dynamical gauge field ignored by [10] is part of the explanation for the failure of the field theory to accurately describe the thermodynamics of the spinon BEC [11].

However, Scammell and Sushkov have identified a useful signature of deconfined spinons: the anomalous specific heat arising from a gapless quadratic mode. The gas of deconfined spinons appearing at higher temperatures is also expected to have such a mode [11, 13] and so we can also expect the spinon gas to exhibit anomalous temperature dependence. The presence of this anomalous temperature dependence is confirmed by our QMC results, thus providing the direct numerical evidence for that there is indeed a gas of deconfined spinons at the deconfined quantum critical point j_c.

Building on this preliminary report, Iaizzi, Sandvik, Scammell, and Sushkov are now preparing an improved analysis combining numerical results with refined theoretical calculations for both the spinon BEC and spinon gas accounting for the presence of a previously ignored dynamical gauge field [13].

The results we present here represent just the first steps of exploring the magnetic field effects in the J-Q model and constructing the full $q - h - T$ phase diagram. With the inclusion of the field, q_c becomes complex point at the intersection of several regimes: Néel antiferromagnet, canted antiferromagnet, valence-bond solid, a BKT transition, and deconfined quantum criticality. Further study is required to develop an understanding of how the field influences the Néel-VBS transition, determine $q_c(h)$, assess the possibility of field-induced phase coexistence between Néel and VBS order, and establish the boundaries for the BKT phase as a function of q and h.

Author's note: Substantial thanks are owed to my collaborators Harley D. Scammell and Oleg P. Sushkov for the interpretation of the data presented here. They suggested that the region above the BKT transition is a gas of deconfined spinons and that the unaccounted-for gauge field may explain why the spinon BEC does not produce the predicted anomalous temperature.

References

1. T. Senthil, A. Vishwanath, L. Balents, S. Sachdev, M.P.A. Fisher, Science **303**, 1490. http://dx.doi.org/10.1126/science.1091806 (2004)
2. A.W. Sandvik, Phys. Rev. Lett. **98**, 227202 (2007). http://dx.doi.org/10.1103/PhysRevLett.98.227202
3. R.G. Melko, R.K. Kaul, Phys. Rev. Lett. **100**, 017203 (2008). http://dx.doi.org/10.1103/PhysRevLett.100.017203
4. A.W. Sandvik, Phys. Rev. Lett. **104**, 177201 (2010). http://dx.doi.org/10.1103/PhysRevLett.104.177201

5. A.W. Sandvik, V.N. Kotov, O.P. Sushkov, Phys. Rev. Lett. **106**, 207203 (2011). http://dx.doi.org/10.1103/PhysRevLett.106.207203
6. H. Suwa, A. Sen, A.W. Sandvik, Phys. Rev. B **94**, 144416 (2016). http://dx.doi.org/10.1103/PhysRevB.94.144416
7. H. Shao, W. Guo, A.W. Sandvik, Science **352**, 213 (2016). http://dx.doi.org/10.1126/science.aad5007
8. M. Levin, T. Senthil, Phys. Rev. B **70**, 220403 (2004). http://dx.doi.org/10.1103/PhysRevB.70.220403
9. T. Sulejmanpasic, H. Shao, A.W. Sandvik, M. Ünsal, Phys. Rev. Lett. **119**, 091601 (2017). http://dx.doi.org/10.1103/PhysRevLett.119.091601
10. H.D. Scammell, O.P. Sushkov, Phys. Rev. Lett. **114**, 055702 (2015). http://dx.doi.org/10.1103/PhysRevLett.114.055702
11. H.D. Scammell, O.P. Sushkov, private communication (2018)
12. J.M. Kosterlitz, D.J. Thouless, J. Phys. C Solid State Phys. **5**, L124 (1972). http://stacks.iop.org/0022-3719/5/i=11/a=002
13. A. Iaizzi, H.D. Scammell, O.P. Sushkov, A.W. Sandvik, Direct numerical observation of Bose-Einstein Condensation of deconfined spinons (2018), in preparation
14. P.W. Anderson, Phys. Rev. **86**, 694 (1952). http://dx.doi.org/10.1103/PhysRev.86.694
15. P. Sengupta, A.W. Sandvik, D.K. Campbell, Phys. Rev. B **65**, 155113 (2002). http://dx.doi.org/10.1103/PhysRevB.65.155113
16. F.D.M. Haldane, Phys. Rev. B **25**, 4925 (1982). http://dx.doi.org/10.1103/PhysRevB.25.4925
17. C.K. Majumdar, D.K. Ghosh, J. Math. Phys. **10**, 1388 (1969). http://dx.doi.org/10.1063/1.1664978
18. C.K. Majumdar, D.K. Ghosh, J. Math. Phys. **10**, 1399 (1969). http://dx.doi.org/10.1063/1.1664979
19. R.B. Laughlin, Science **303**, 1475 (2004). http://dx.doi.org/10.1126/science.1095266
20. T. Senthil, L. Balents, S. Sachdev, A. Vishwanath, M.P.A. Fisher, Phys. Rev. B **70**, 144407 (2004). http://dx.doi.org/10.1103/PhysRevB.70.144407
21. J. Lou, A.W. Sandvik, N. Kawashima, Phys. Rev. B **80**, 180414 (2009). http://dx.doi.org/10.1103/PhysRevB.80.180414
22. S. Jin, A.W. Sandvik, Phys. Rev. B **87**, 180404 (2013). http://dx.doi.org/10.1103/PhysRevB.87.180404
23. Y. Tang, A.W. Sandvik, Phys. Rev. Lett. **110**, 217213 (2013). http://dx.doi.org/10.1103/PhysRevLett.110.217213
24. E. Fradkin, D.A. Huse, R. Moessner, V. Oganesyan, S.L. Sondhi, Phys. Rev. B **69**, 224415 (2004). http://dx.doi.org/10.1103/PhysRevB.69.224415
25. A.B. Kuklov, M. Matsumoto, N.V. Prokof'ev, B.V. Svistunov, M. Troyer, Phys. Rev. Lett. **101**, 050405 (2008). http://dx.doi.org/10.1103/PhysRevLett.101.050405
26. F.-J. Jiang, M. Nyfeler, S. Chandrasekharan, U.-J. Wiese, J. Stat. Mech. Theory Exp. **2008**, P02009 (2008). http://stacks.iop.org/1742-5468/2008/i=02/a=P02009
27. A. Iaizzi, K. Damle, A.W. Sandvik, Phys. Rev. B **95**, 174436 (2017). http://dx.doi.org/10.1103/PhysRevB.95.174436
28. N.D. Mermin, H. Wagner, Phys. Rev. Lett. **17**, 1133 (1966). http://dx.doi.org/10.1103/PhysRevLett.17.1133
29. A.W. Sandvik, J. Kurkijärvi, Phys. Rev. B **43**, 5950 (1991). http://dx.doi.org/10.1103/PhysRevB.43.5950
30. A.W. Sandvik, in *American Institute of Physics Conference Series*, ed. by A. Avella, F. Mancini, vol. 1297 (2010), pp. 135–338. http://arxiv.org/abs/1101.3281. http://dx.doi.org/10.1063/1.3518900
31. O.F. Syljuåsen, A.W. Sandvik, Phys. Rev. E **66**, 046701 (2002). http://dx.doi.org/10.1103/PhysRevE.66.046701
32. A.W. Sandvik, Phys. Rev. B **66**, 024418 (2002). http://dx.doi.org/10.1103/PhysRevB.66.024418
33. K. Hukushima, K. Nemoto, J. Phys. Soc. Jpn. **65**, 1604 (1996). http://dx.doi.org/10.1143/JPSJ.65.1604

34. A. Iaizzi, K. Damle, A.W. Sandvik, Phys. Rev. B **98**, 064405 (2018). http://dx.doi.org/10.1103/PhysRevB.98.064405
35. T. Einarsson, H.J. Schulz, Phys. Rev. B **51**, 6151 (1995). http://dx.doi.org/10.1103/PhysRevB.51.6151
36. A.L. Chernyshev, M.E. Zhitomirsky, Phys. Rev. B **79**, 174402 (2009). http://dx.doi.org/10.1103/PhysRevB.79.174402
37. D.R. Nelson, J.M. Kosterlitz, Phys. Rev. Lett. **39**, 1201 (1977). http://dx.doi.org/10.1103/PhysRevLett.39.1201
38. Y.-D. Hsieh, Y.-J. Kao, A.W. Sandvik, J. Stat. Mech. Theory Exp. **2013**, P09001 (2013). http://dx.doi.org/10.1088/1742-5468/2013/09/P09001
39. D.P. Landau, K. Binder, Phys. Rev. B **24**, 1391 (1981). http://dx.doi.org/10.1103/PhysRevB.24.1391
40. A.S.T. Pires, Phys. Rev. B **50**, 9592 (1994). http://dx.doi.org/10.1103/PhysRevB.50.9592
41. A.W. Sandvik, Phys. Rev. B **59**, R14157 (1999). http://dx.doi.org/10.1103/PhysRevB.59.R14157
42. A. Cuccoli, T. Roscilde, R. Vaia, P. Verrucchi, Phys. Rev. B **68**, 060402 (2003). http://dx.doi.org/10.1103/PhysRevB.68.060402
43. A. Cuccoli, T. Roscilde, R. Vaia, P. Verrucchi, J. Magn. Magn. Mater. **272–276, Part 2**, 884 (2004). http://dx.doi.org/10.1016/j.jmmm.2003.12.605
44. L. Baranová, A. Orendáčová, E. Čižmár, R. Tarasenko, V. Tkáč, M. Orendáč, A. Feher, J. Magn. Magn. Mater. **404**, 53 (2016). http://dx.doi.org/10.1016/j.jmmm.2015.12.025
45. N. Barbero, T. Shiroka, C.P. Landee, M. Pikulski, H.-R. Ott, J. Mesot, Phys. Rev. B **93**, 054425 (2016). http://dx.doi.org/10.1103/PhysRevB.93.054425
46. A. Cuccoli, T. Roscilde, V. Tognetti, R. Vaia, P. Verrucchi, Phys. Rev. B **67**, 104414 (2003). http://dx.doi.org/10.1103/PhysRevB.67.104414
47. A. Pires, Solid State Commun. **112**, 705 (1999). http://dx.doi.org/10.1016/S0038-1098(99)00405-6
48. R.K. Kaul, R.G. Melko, Phys. Rev. B **78**, 014417 (2008). http://dx.doi.org/10.1103/PhysRevB.78.014417
49. R.K. Pathria, P.D. Beale, *Statistical Mechanics*, 3rd edn. (Elsevier, New York, 2011)

Chapter 5
Methods

This dissertation is based almost entirely on numerical methods. Chief among these is the stochastic series expansion (SSE) quantum Monte Carlo (QMC) method [1–3], although I have also used Lanczos exact diagonalization for small systems, verification and other special reasons.

Outline I will describe exact diagonalization methods briefly in Sect. 5.1. In the rest of this chapter I will develop the quantum Monte Carlo methods that I have used in this dissertation. I have attempted to make this chapter a pedagogically useful guide for the reader interested in replicating or building upon this work. I begin by describing the foundations of classical Monte Carlo in Sect. 5.2. In Sect. 5.3 I derive the stochastic series expansion formulation of quantum Monte Carlo. I then describe applications of this method to the Heisenberg model (Sect. 5.4), the J-Q model (Sect. 5.5), and the Heisenberg model in an external field (Sect. 5.6) where I also explain directed loop updates. In Sect. 5.7, I synthesize the previous three sections, describing the QMC method used here for the J-Q model in an external field. I describe the supplementary techniques quantum replica exchange and β-doubling in Sect. 5.8 and finish with a brief discussion of random number generators in Sect. 5.9.

5.1 Exact Diagonalization

Exact diagonalization is the most direct approach to discrete quantum systems, but it is limited to extremely small sizes. The Hilbert space of quantum Hamiltonians grows exponentially with system size; for an $S = \frac{1}{2}$ chain, the Hilbert space is 2^L. This can be partially alleviated with the use of symmetries to block-diagonalize the Hamiltonian so that each block can be diagonalized separately. For example, all the Hamiltonians used here commute with the total S^z operator, ($[S^z, H] = 0$), so eigenstates of H will also be eigenstates of S^z. There will be no nonzero matrix

A. Iaizzi, *Magnetic Field Effects in Low-Dimensional Quantum Magnets*,
Springer Theses, https://doi.org/10.1007/978-3-030-01803-0_5

elements of H connecting states with different S^z eigenvalues (given by m_z), i.e., $\langle \psi, m_z = 1|H|\psi', m_z = 3\rangle = 0$. Breaking H up into blocks like this can provide a significant speedup and allow us to solve larger systems, but in the end the Hilbert space still grows exponentially, so system sizes are still limited.

Full diagonalization returns all eigenvector and eigenvalues of the Hamiltonian. We can reach larger systems using Lanczos diagonalization, which determines the ground state and first few excited states by using repeated action of the Hamiltonian to project out the most extreme eigenvalues. Even with all of these speedups implemented, the Lanczos exact diagonalization is still limited to extremely small systems (the world record is currently in the neighborhood of a 40 site system).[1] The computational expense quantum Monte Carlo, on the other hand, grows as polynomial of the system size. I used exact diagonalization primarily for studying small systems and validating my quantum Monte Carlo programs. I will not describe these methods in detail here because the methods I have used are all quite standard. A detailed explanation of exact diagonalization as it applies to spin systems, including both full and Lanczos approaches, implementation of symmetries, and pseudocode, can be found in Ch. 4 of [3]. There is also an excellent package written by my colleagues Phillip Weinberg and Marin Bukov called QuSpin, which does exact diagonalization of a user-defined spin Hamiltonian with all applicable symmetries from a Python interface [4].[2]

5.2 Monte Carlo

One of the earliest papers on Monte Carlo [5] defined the method thusly:

> The Monte Carlo method is an iterative stochastic procedure, consistent with a defining relation for some function f, which allows an estimate of another function of f without completely determining f. (Metropolis and Ulam, [5])

This definition is accurate, but perhaps not especially useful. The key word in this definition is *stochastically*: Monte Carlo describes a wide range of numerical techniques that use random numbers to solve problems. Monte Carlo can be used to evaluate quantities that involve obvious probabilities, the most famous of which among nonscientists might be Nate Silver's election-prediction algorithm hosted on fivethirtyeight.com. But Monte Carlo can also be used to evaluate quantities that have nothing to do with probability. The canonical first example is numerically estimating π by finding the area of the unit circle. We begin with an integral:

---

[1]A side effect of the exponential growth is that only a factor of two separates the size of a system that can be solved in an hour by a simple code running on a laptop and the largest that can be solved using a state-of-the-art code on a supercomputer.

[2]Usage and installation procedures for QuSpin can be found in [4]. QuSpin can be installed from the package manager Anaconda or from Github: https://github.com/weinbe58/QuSpin.

$$I = \int_{-1}^{1}\int_{-1}^{1} f(x, y)dxdy = \pi \tag{5.1}$$

where $f(x, y)$ is defined

$$f(x, y) = \begin{cases} 1, & \text{if } \sqrt{x^2 + y^2} \le 1 \\ 0, & \text{otherwise.} \end{cases} \tag{5.2}$$

To evaluate this integral using Monte Carlo, we draw pairs of random numbers on a uniform interval $x_i = [0, 1]$, $y_i = [0, 1]$ and compute the sum:

$$\langle I \rangle = \frac{1}{N} \sum_{i}^{N} f(x_i, y_i) = \frac{N_{\text{inside}}}{N} \tag{5.3}$$

where the ratio of the number of pairs (x_i, y_i) falling inside the unit circle to the total number of pairs drawn will converge to π in the limit $N \to \infty$. The statistical error in $\langle I \rangle$ for finite N will be proportional to $1/\sqrt{N}$.

This procedure seems rather contrived: why not just use direct numerical integration?[3] Numerical integration (using Romberg integration, for example) breaks the integral into steps Δ, and evaluates the integral by summing the area of these small pieces, introducing an error $\delta \propto \mathcal{O}(\Delta^2)$. In 1D, the step size, $\Delta \propto 1/N$ is inversely proportional to the number of steps, N, so the discretization error is then $\propto \mathcal{O}(1/N^2)$. The number of points required, N, is a rough measure of the computational expense of a calculation. Therefore direct numerical integration is superior to Monte Carlo integration in one dimension. In higher dimensions, however, the number of points required grows rapidly $N = \Delta^{-d}$. In order to keep a constant discretization error, δ, the computational costs therefore scales like $N \propto \left(\frac{1}{\delta}\right)^{d/2}$. The cost of direct integration is *exponential* in d. This may seem like a minor consideration for double and triple integration, but quantum Monte Carlo programs routinely evaluate sums of $d = \mathcal{O}(10^5)$ or higher. The statistical error in Monte Carlo integration does not depend on dimension at all, it is always $\mathcal{O}(N^{-1/2})$. Monte Carlo produces no discretization error,[4] and when done correctly, the statistical error in Monte Carlo simulations is Gaussian-distributed with known error bars. Monte Carlo integration is therefore extremely well suited to high-dimensional sums and integrals. In physics Monte Carlo methods generally rely

[3] In the case of evaluating π, this example is especially contrived because there are far more precise specialized procedures for calculating π.

[4] The discretization error is a kind of systematic error (as opposed to statistical or random error) which is not Gaussian and does not have well-defined error bars.

on mapping a problem onto a high-dimensional sum which is then evaluated by stochastic sampling.

Let us consider a more practical example: say you want to know the expectation value of some quantity, $\hat{O}$, in a classical system governed by the Boltzmann distribution. The expectation value $\langle \hat{O} \rangle$ can be written in terms of the Boltzmann weights:

$$\langle \hat{O} \rangle = \frac{1}{Z} \sum_{x} \hat{O}_x e^{-\beta E_x} \tag{5.4}$$

where each x represents a configuration of the system, for example, spin states in the 2D Ising model and Z is the partition function:

$$Z = \sum_{i=1}^{N} e^{-\beta E_{x(i)}}. \tag{5.5}$$

Equation (5.4) is a well-defined sum that could—in principle—be evaluated directly, but in practice this is impossible. In the case of the 2D Ising model, the sum over x in Eq. (5.4) is every possible combination of up and down spins on each of the L^2 sites—an L^2-dimensional sum with a total of $2^{L^2} \approx 10^{3010}$ terms for $L = 100$. Rather than evaluate all these terms, we will use random sampling Monte Carlo. We do not need to sum over *all* the configurations, we can instead just average over a large random sample of them and we should get an accurate result. Let us stick to the Ising model for our example. To evaluate the expectation value $\langle \hat{O} \rangle$ with random sampling Monte Carlo we generate N random Ising model configurations (easy enough) and perform an average weighted with the Boltzmann distribution:

$$\langle \hat{O} \rangle = \frac{1}{Z} \sum_{i=1}^{N} \hat{O}_{x(i)} e^{-\beta E_{x(i)}} \tag{5.6}$$

Note here that our sample will never cover a macroscopic fraction of the total configuration space. Even if we draw a billion (10^9) random samples, they would comprise an incomprehensibly minuscule fraction of the 10^{3010} states in the full configuration space.

5.2.1 *Importance Sampling*

The random sampling approach works, but it has a huge downside: you spend a lot of time sampling unimportant points in configuration space, like high-energy configurations that contribute very little to your expectation value (because their weights are small). An improved (and almost universally used) version of Monte

Carlo is called *importance sampling*. In importance sampling, configurations are drawn directly from the relevant probability distribution. $\langle \hat{O} \rangle$ can be calculated by drawing N Ising configurations from the Boltzmann distribution and average them directly:

$$\langle \hat{O} \rangle = \frac{1}{N} \sum_{i=1}^{N} \hat{O}_{x(i)} \tag{5.7}$$

Here the Boltzmann weights do not appear in the average because the configurations $\{x\}$ are already drawn from the correct probability distribution $P \propto e^{-\beta E_x}$. The likelihood of these configurations is already accounted for by drawing them from the distribution. Importance sampling "does what it says on the tin": it focuses on sampling the most important points in the distribution, and it therefore requires fewer samples to achieve the same level of precision,[5] and it is especially valuable for sharply peaked distributions (like the Boltzmann distribution) and "sparse" sums where most of the terms are zero (as will be the case for the stochastic series expansion quantum Monte Carlo method used in this dissertation).

5.2.2 *What Is a Markov Process?*

The challenge in importance sampling is to come up with an efficient way of generating configurations drawn from the appropriate distribution. This is usually accomplished using a *Markov Process*, which works like this:

1. Choose a starting configuration, x (at random).
2. Propose a change to that configuration $x \to x'$
3. Accept that change with some probability $A(x \to x')$
4. Return to step 2, repeat.

Iterating this procedure produces a time series of configurations called *Markov chain* and this type of Monte Carlo is called Markov chain Monte Carlo.[6] Although the Markov chain is a time series (in simulation time) the dynamics of the updates have nothing to do with physical time or physical dynamics. Another important aspect of this procedure is that the transition probability depends *only* on the present configuration, and not on previous configurations: "The future depends on the past, but only through the present."[7]

[5]The uniformly weighted average is the most "efficient" estimator for the mean (i.e., it has the lowest variance) [6, p. 135].

[6]Often abbreviated MCMC.

[7]Said by Lode Pollet during his lecture on QMC at the Arnold Sommerfeld Center at LMU Munich as part of the Arnold Sommerfeld School on 13 September 2017.

A Markov process consists of a set of transition probabilities $P(x \to x')$ that maintains a *stationary distribution* such that the probability of occupying a state x, given by $\pi(x)$ is constant over time. A set of transition probabilities is a Markov process if it meets all the following conditions:

1. There exists a *stationary distribution*, $\pi(x)$, where there is no net probability flow between states (i.e., the *global balance* condition).
2. The process is aperiodic: not repeating (although it may visit the same state multiple times).
3. The process must be ergodic, i.e., any state x' is accessible from any other state x through some finite number of steps along the chain.

To maintain a stationary state, the updates must obey the *global balance* condition:

$$\sum_x \pi(x)P(x \to x') = \sum_{x'} \pi(x')P(x' \to x) \tag{5.8}$$

where $\pi(x)$ the probability of occupying state x and $P(x \to x')$ is the probability of transitioning from state x to x' given that you are already in state x. Simply stated: there is no net probability flux to or from any state.

Determining if global balance is satisfied is typically difficult, so most Markov chain Monte Carlo schemes instead use the *detailed balance* condition:

$$\pi(x)P(x \to x') = \pi(x')P(x' \to x) \tag{5.9}$$

Detailed balance requires no net flow between any pair of states x, x'. This is a stricter condition (i.e., if detailed balance is satisfied, global balance is always satisfied), but it is usually simpler to enforce. There are Monte Carlo schemes that utilize global balance (see [7]), but they are the exception rather than the norm.

5.2.3 The Metropolis–Hastings Algorithm

There are many solutions to the detailed balance condition,[8] one of the most commonly used is the Metropolis–Hastings[9] algorithm originally developed by Arianna Wright Rosenbluth, Marshall N. Rosenbluth, Augusta H. Teller, and Edward Teller [8] and later generalized by W.K. Hastings [12]. Today Monte Carlo

[8]In fact, there are infinitely many solutions to the detailed balance condition.

[9]This algorithm is usually referred to simply as the Metropolis Algorithm although perhaps it should be called the Rosenbluth or Rosenbluth–Teller Algorithm. Metropolis was first author on the original paper [8], but according to Marshall Rosenbluth [9, 10] Metropolis was merely the head of the computer lab and made no scientific contribution to the paper. In Metropolis' memoirs, he makes no claim to have invented the algorithm either [11]. See Sect. 1.2.2 for a more complete discussion.

is difficult to separate from the Metropolis Algorithm. The Metropolis Algorithm was the first practical and recognizably modern iteration of Monte Carlo that used importance sampling (instead of random sampling). The first functional Metropolis Algorithm code was written entirely by Arianna Wright Rosenbluth [9, 11, 13]. In the remainder of this section I will derive the Metropolis–Hastings Algorithm and discuss technical aspects of its invention; for a discussion of the scientific significance of the Metropolis Algorithm and story of its creation I refer the reader to Sect. 1.2.2.

We begin with the detailed balance condition from Eq. (5.9), and replace the probabilities $\pi(x)$ with unnormalized weights $W(x)$

$$\frac{W(x)}{\mathcal{N}} P(x \to x') = \frac{W(x')}{\mathcal{N}} P(x' \to x)$$

When Monte Carlo is used for statistical physics simulations, these weights will generally depend on the energy, for example, they may be the Boltzmann weights $e^{-\beta E_i}$. This normalization factor is typically difficult to compute (since it involves a huge multidimensional sum). Fortunately, we can avoid calculating it by rearranging to cancel $\mathcal{N}$:

$$\frac{P(x \to x')}{P(x' \to x)} = \frac{W(x')}{W(x)} \tag{5.10}$$

Next we separate the transition probability $P(x \to x')$ into two parts: proposal and acceptance. The probability of proposing a transition from x to x' given x, $g(x \to x')$, and the probability of accepting the proposed change, $A(x \to x')$.

$$P(x \to x') = g(x \to x')A(x \to x') \tag{5.11}$$

Both the proposal probability, g, and the acceptance probability, A, will depend on the type of update that is being used.[10] We can use this to rewrite the detailed balance condition:

$$\begin{aligned} \frac{g(x \to x')A(x \to x')}{g(x' \to x)A(x' \to x)} &= \frac{W(x')}{W(x)} \\ \frac{A(x \to x')}{A(x' \to x)} &= \frac{W(x')g(x' \to x)}{W(x)g(x \to x')} \end{aligned} \tag{5.12}$$

So far we have merely restated the detailed balance condition in a slightly different form. Now we must choose $A(x \to x')$ such that Eq. (5.12) is satisfied.

[10] For example, in an implementation of the Metropolis Algorithm for the Ising model, one selects a spin at random and proposes to flip that spin; the probability of proposing this change is therefore $1/N$ and the probability of proposing the reverse change is also $1/N$ [3, Sec. 3.2].

The Metropolis–Hastings choice for A is given by

$$A(x \to x') = \min\left(1, \frac{W(x')g(x' \to x)}{W(x)g(x \to x')}\right) \tag{5.13}$$

As a gross simplification: in the Metropolis Algorithm, you *always* accept a change if it lowers the energy, and *sometimes* accept a change if it increases the energy. There are other ways of satisfying the detailed balance condition, but the Metropolis–Hastings Algorithm is by far the most common. Even the sophisticated quantum Monte Carlo algorithms used for this work rely on the Metropolis–Hastings Algorithm for some of the updates and rely generally on the approach to Monte Carlo that was first proposed along with the Metropolis Algorithm (see Sect. 1.2.2).

5.2.4 Practical Considerations: Autocorrelations, Binning, Error Bars, and Equilibration

Autocorrelations Markov chain Monte Carlo efficiently produces configurations drawn from the appropriate distribution, but at a cost: consecutive configurations are no longer independent. Each change leaves most of the configuration unaltered and changes are often rejected, in which case the configuration is completely unchanged. The correlations between states within a Markov chain are called *autocorrelations*. Autocorrelations decay exponentially with the number of Monte Carlo steps (attempted updates). The exponential decay constant is known as the *autocorrelation* time (here time refers to simulation time).

Measurement We will refer to a Monte Carlo sweep as a given number of attempted updates n_s. The number of attempted updates in each sweep is usually set to scale with the system size; for example, in the 2D Ising model we might set $n_s = L^2$. After each sweep, we will perform a *measurement*. Even after n_s attempted updates, the consecutive measurements will still be correlated. There is a tradeoff here worth mentioning: the more frequent the measurements, the more correlated consecutive measurements will be. This is not itself a problem, but some measurements can be computationally expensive (especially correlation functions), so measuring too frequently is a waste of time. On the other hand, measuring too infrequently is throwing away valuable information. Although the right balance is difficult to define in advance, the time spent performing measurements should not dominate the computational cost of a simulation.

Binning Over the course of a simulation, many sweeps are performed, often at least 10^6, but there is no need to write every one of these measurements out to file. Writing too frequently to disk is a waste of CPU time and disk space, especially since consecutive measurements are often highly correlated. Instead, our Monte

Carlo program will bin this data, putting n_{steps} measurements in each bin and writing just the averages for each bin out to file. This dramatically reduces the size of the output files with minimal loss of information. Ideally the bin size should be longer than the autocorrelation time, so that the data in each bin is uncorrelated with the preceding bin. Another advantage of binning is that we can take advantage of the *central limit theorem*, which states that the distribution of the averages, y of any quantity x

$$y_j = \frac{1}{n_{\text{steps}}} \sum_i^{n_{\text{steps}}} x_i \tag{5.14}$$

will be normally distributed regardless of the distribution of the original quantity x provided that n_{steps} is sufficiently large [6, Sec. 5.9]. The expectation value of any quantity, q, is then just the average over the n_{bins} bins:

$$\langle q \rangle = \frac{1}{n_{\text{bins}}} \sum_i^{n_{\text{bins}}} q_i \tag{5.15}$$

The bin values q_i will be Gaussian-distributed with standard deviation:

$$\sigma(q_i) = \sqrt{\frac{1}{n_{\text{bins}} - 1} \sum_i^{n_{\text{bins}}} (q_i - \langle q \rangle)^2} \tag{5.16}$$

The standard deviation of the bin values q_i will not depend on the number of bins, only on the size of each bin, n_{steps}, and should be proportional to $\frac{1}{\sqrt{n_{\text{steps}}}}$. To determine the error bars we use the standard deviation of the mean of q_i, $\sigma(\langle q \rangle)$:

$$\sigma(\langle q \rangle) = \frac{\sigma(q_i)}{\sqrt{n_{\text{bins}}}} = \sqrt{\frac{1}{n_{\text{bins}}(n_{\text{bins}} - 1)} \sum_i^{n_{\text{bins}}} (q_i - \langle q \rangle)^2} \tag{5.17}$$

Earlier we mentioned that the bins must be larger than the autocorrelation time, this is essential for calculating accurate error bars. Bins that are too small will cause an *underestimate* of the error bars. The error bars for $\langle q \rangle$ should depend only on the total number of Monte Carlo sweeps performed, $n_{\text{bins}} \times n_{\text{steps}}$. Therefore, rebinning (changing the size of the bins after the fact) should not affect $\sigma(\langle q \rangle)$. If we double the bin size and halve the number of bins, we arrive at new bins q_i':

$$q_i' = \sum_{j=1,3,5...}^{n_{\text{bins}}-1} \frac{q_j + q_{j+1}}{2} \tag{5.18}$$

The error bars calculated using the rebinned data, $\sigma(\langle q' \rangle)$, should be about the same as the original error bars, $\sigma(\langle q \rangle)$. If they are not, the data in consecutive bins are still correlated and the bins are too short (n_{steps} is too small). This problem can be solved by repeatedly rebinning until the error bars converge to a constant value. You can convince yourself of this by considering the most extreme case where the configuration does not change at all, then each measurement is the same and the standard deviation (and therefore the error bars) goes to zero.

Optimal Bin Size Ideally, bins should be long enough that $n_{\text{steps}} \gg \tau_{\text{AC}}$ (the autocorrelation time). But τ_{AC} is not known in advance, and if the bins are too small, that can always be resolved by *rebinning*. Here there are some practical issues to consider as well. Writing to disk is glacially slow, so writing out to file frequently (more than once per second) will slow a program down dramatically. In theory, there is no problem with bins that are too long, but a bin is a quantum of simulation progress. If bins are too long, becomes difficult to tell if a simulation is running properly and if the program crashes (or is stopped early) all the progress since the last bin written to disk is lost. The actual amount of wall time required to complete a bin will vary based on factors like system size, temperature, and proximity to critical points, but it should not be less than 1 min and no more than 24 h. Queuing systems (like on the SCC) tend to be designed for jobs that last between 1 and 24 h, so if a simulation can make substantial progress in that time, that is ideal. It is also desirable to have at least ten bins for the purposes of calculating error bars, and more to enable rebinning to check for autocorrelations. In practice, I have found that I prefer no fewer than 20 bins, and no more than 100 (for a first run). If the error bars are too large, one can always run the simulation again (with a different random seed), appending additional bins to the same files to accumulate more data.

Equilibration The initial configuration for a Monte Carlo simulation is typically a randomized configuration that corresponds to an infinite-temperature state. It will take some time for the updates to change this configuration enough to arrive at an equilibrium configuration. This process is referred to as equilibration. To account for this, we must wait until enough Monte Carlo steps have passed before we begin collecting data. Typically the equilibration time is one bin's worth of Monte Carlo sweeps. For some systems, this process of equilibration can be extremely slow since the sudden change from an infinite-temperature configuration to low-temperature dynamics is equivalent to a quench. In those cases, enhanced equilibration procedures can be employed; one of the most common of these techniques is simulated annealing, where the temperature is slowly reduced to the desired value over a period of time.

5.3 Quantum Monte Carlo: The Stochastic Series Expansion

The stochastic series expansion (SSE) quantum Monte Carlo (QMC) method is the numerical workhorse of this study. Monte Carlo is a method of evaluating sums or integrals using stochastic sampling with classical probabilities. The term *quantum* Monte Carlo describes a family of techniques for mapping a quantum problem onto a classical problem. Once the analogous classical problem has been defined, it is evaluated using classical Monte Carlo. The stochastic series expansion is one method for performing this mapping. Other examples include projector QMC, continuous time QMC, determinant QMC (for fermions), diagrammatic QMC, and diffusion QMC. The computational expense stochastic series expansion is polynomial in system size, L, and inverse temperature, $\beta = 1/T$, scaling like βL^d. SSE can be used to access zero-temperature properties by exploiting the gap to the first excited state that appears in finite-size systems. This gap is typically $\Delta \propto 1/L$, so zero-temperature properties appear when $T \ll \Delta$, i.e., $\beta \gg L$. Therefore, SSE can access zero-temperature properties at a computational cost that scales $\propto L^{d+1}$.

Outline In this section I will derive the stochastic series expansion formulation of quantum Monte Carlo, and in Sects. 5.4 to 5.7 I will describe applications, starting with the Heisenberg model, then moving on to describe how the four-spin Q term is implemented, followed by the Heisenberg model in an external magnetic field. Finally, in Sect. 5.7, I synthesize the work of the intervening sections to apply SSE to the J-Q-h model. The formulations described in Sects. 5.4 to 5.6 are not new, but I hope for this chapter to be pedagogically useful, so I have used them to build toward the J-Q-h model application (which is new), which is the core of this dissertation.

5.3.1 Formalism

We can extract expectation values of finite-temperature properties of a classical system by evaluating the (canonical ensemble) partition function at inverse temperature $\beta \equiv 1/T$:

$$Z = \sum_i e^{-\beta E_i} \tag{5.19}$$

where the sum over i indicates a sum over all possible states. The quantum analog of the partition function is the trace over the *density matrix*

$$Z = Tr\{\rho\} = Tr\{e^{-\beta H}\}. \tag{5.20}$$

The Hamiltonian, H, is an operator that is typically so large that it is impossible to exponentiate. For example, in a spin-half chain H is a $2^L \times 2^L$ matrix.

To evaluate this partition function using Monte Carlo, we must convert the partition function into a high-dimensional sum. First we expand in a Taylor series:

$$Z = \sum_{\alpha_0} \left\langle \alpha_0 \left| e^{-\beta H} \right| \alpha_0 \right\rangle = \sum_{\alpha_0} \sum_{n} \frac{\beta^n}{n!} \left\langle \alpha_0 \left| (-H)^n \right| \alpha_0 \right\rangle \tag{5.21}$$

Each term in this Taylor series is still too hard to evaluate since each involves taking a power of an exponentially large Hamiltonian. For now, let us consider just one term in this expansion, $\langle \alpha_0 \mid (-H)^n \mid \alpha_0 \rangle$, and insert a complete set of states, $\sum_{\alpha_i} |\alpha_i\rangle \langle \alpha_i|$ between each instance of H:

$$\left\langle \alpha_0 \left| (-H)^n \right| \alpha_0 \right\rangle = \sum_{\alpha_1} \cdots \sum_{\alpha_{n-1}} (-1)^n \langle \alpha_0 | H | \alpha_1 \rangle \langle \alpha_1 | H | \alpha_2 \rangle \ldots \langle \alpha_{n-1} | H | \alpha_0 \rangle \tag{5.22}$$

Expanding in terms of $n-1$ complete sets of states greatly expands the number of terms in the summation, but instead of exponentiating H, or even raising it to a power, we now only need to know matrix elements of the Hamiltonian between specific basis states, for example, $\langle \alpha_3 | H | \alpha_4 \rangle$. Typical Hamiltonians in correlated electron systems (like the Heisenberg or Hubbard model) are usually sums of local interactions, so individual matrix elements between states are typically simple and easy to evaluate.[11]

Now plugging Eq. (5.22) into Eq. (5.21) we find a new expression for Z, the classical sum that we will evaluate using Monte Carlo:

$$Z = \sum_{n} \sum_{\{\alpha\}} \frac{(-\beta)^n}{n!} \langle \alpha_0 | H | \alpha_1 \rangle \langle \alpha_1 | H | \alpha_2 \rangle \ldots \langle \alpha_{n-1} | H | \alpha_0 \rangle \tag{5.23}$$

where the sum over $\{\alpha\}$ is a sum over all possible combinations of the α_i's, replacing the $n-1$ sums over $\alpha_0, \alpha_1, \alpha_2 \ldots \alpha_{n-1}$ in Eq. (5.22). Equation (5.23) is a high-dimensional sum in a large space—a natural candidate for Monte Carlo. Moreover, this case is especially well suited to *importance sampling* since almost all of the terms in the sum will be zero. H is extremely sparse, so a matrix element connecting any two states is very likely to be zero-valued.

It is worth making note here of another feature; we will be computing transition probabilities using Eq. (5.23) as the weight of each configuration. There is a key difference here from the classical case: every term in the classical partition function [Eq. (5.19)] is positive and real. Therefore, it is trivial to guarantee that the Monte Carlo weights assigned to each configuration are also positive and real. The same cannot be said for the terms in Eq. (5.23); generically, some of them *will* be negative or even possibly imaginary; this is known as the *sign problem*. Avoiding the sign problem is one of the central challenges and limitations of quantum Monte Carlo. The sign problem is almost always present in fermionic systems (except where there is a mapping onto an effective bosonic model) and it is also usually present

[11]For example, in the Heisenberg model, after some simple transformations there are only four nonzero local matrix elements which all have the same value, Eq. (5.35).

in frustrated spin systems. The J-Q model was designed to emulate some of the features of frustrated systems in a sign-problem-free manner.

Each Monte Carlo "configuration" is a full set of states $\alpha_0, \alpha_1, \ldots \alpha_{n-1}$ along with the operators connecting them. This configuration has an extra dimension compared to the original problem. In this case, the extra dimension corresponds roughly to imaginary time. We are expanding the quantity $e^{-\beta H}$, which has the same form as the time evolution operator:

$$e^{-\beta H} \Rightarrow e^{-itH} = e^{\tau H} = U(t) \tag{5.24}$$

where τ is imaginary time $t = -i\tau$. Thus, SSE is often referred to as an "imaginary time expansion," and the α_i's can be thought of as a time series. Each set of intermediate states $\{\alpha\}$ constitutes a "path" in configuration space, so the sum of $\{\alpha\}$ is a sum over paths, thus this expansion also corresponds to a path integral. Here the "paths" are the sets of many-body states the spin system propagates through rather than locations in real space (i.e., paths through configuration space). A generic property of quantum Monte Carlo methods is this sort of mapping of a d-dimensional quantum problem onto a $(d+1)$-dimensional classical problem.

Our task is now to develop a way to efficiently sample the space of n and $\{\alpha\}$ while ensuring that the terms in the sum [Eq. (5.23)] are all ≥ 0. We will accomplish this using operator-loop updates, a method was first introduced by Sandvik and Kurkijärvi [1] alongside the SSE formulation as a generalization of Handscomb's method.[12] SSE and operator-loop updates [15] have since been expanded using enhanced update schemes such as directed loops [2] and quantum replica exchange [16, 17].

Before we examine our first example, we will add one practical programming consideration. To make inserting and removing operators easier, we will fix the number of complete sets of states to be M, (which we will call the cutoff) and "pad" the product with identity operators. This allows the arrays that store the SSE configuration to be of fixed size, avoiding costly resizing operations. The cutoff M will be larger than the largest order of the expansion (n), so this procedure is exact and does not introduce any error. In practice, M can be determined automatically during the equilibration phase of the simulation by setting $M = \frac{4}{3} n_{\max}$, where $n_{\max}$ is the largest value of n reached so far.[13] After the equilibration phase, the cutoff must be fixed to avoid biasing the simulation.

We can rewrite Eq. (5.23) in terms of M operators O_i. n of these operators O_i are local matrix elements of H and the remaining $M - n$ are identity operators, $\mathbb{I}$:

$$Z = \sum_{n=0}^{M} \sum_{\{\alpha\}} \frac{(-\beta)^n}{n!} \frac{n!(M-n)!}{M!} \langle \alpha_0 | O_0 | \alpha_1 \rangle \langle \alpha_1 | O_1 | \alpha_2 \rangle \ldots \langle \alpha_{M-1} | O_{M-1} | \alpha_0 \rangle$$

[12] A brief history of the development of stochastic series expansion is available in [14].

[13] The exact fraction used here is not important; any number greater than unity will work.

$$Z = \sum_{n=0}^{M} \sum_{\{\alpha\}} \frac{(-\beta)^n (M-n)!}{M!} \langle\alpha_0| O_0 |\alpha_1\rangle \langle\alpha_1| O_1 |\alpha_2\rangle \ldots \langle\alpha_{M-1}| O_{M-1} |\alpha_0\rangle \tag{5.25}$$

To avoid double counting, we must divide by the binomial coefficient $\binom{M}{M-n} = \frac{M!}{n!(M-n)!}$.[14] Finally with Eq. (5.25) we have arrived at the sum that we will directly evaluate using our Monte Carlo program. Each term in Eq. (5.25) is a configuration in our simulation space: a set of $\{\alpha\}$ with M operators O_i (which are completely determined by the $\{\alpha\}$). When calculating transition probabilities we will need to determine the weight[15] of a single configuration $\{\alpha\}$, which is simply the value of the corresponding term in the sum:

$$W(\{\alpha\}) = \frac{(-\beta)^n (M-n)!}{M!} \langle\alpha_0| H |\alpha_1\rangle \langle\alpha_1| H |\alpha_2\rangle \ldots \langle\alpha_{n-1}| H |\alpha_0\rangle \tag{5.26}$$

From the previous section, we now know that there is a short list of values of the matrix elements $\langle\alpha_p|H|\alpha_{p+1}\rangle$, so we can rewrite Eq. (5.26) in terms of a short list of nonzero operators types where there are n_i of operator type i with weight W_i

$$W = \frac{(-\beta)^n (M-n)!}{M!} (W_1)^{n_1} (W_2)^{n_2} (W_3)^{n_3} \ldots \tag{5.27}$$

where $\sum_i n_i = n$. In practice, we will never evaluate the weight itself, but the quotient of the weights of closely related configurations, sidestepping the issue of evaluating the factorials of large numbers in Eq. (5.27).

A brief aside to comment on how far we have come. We began in Eq. (5.19) with an exponentiation of an exponentially large matrix and transformed that problem into the Monte Carlo friendly sum in Eq. (5.25). On its face, Eq. (5.25) is a completely intractable sum—each α_i can take on 2^N values, so there are $(2^N)^M$ terms in this Hamiltonian, for a modest simulation with $N = 100^2$ and $M = \mathcal{O}(10^5)$ there are then $\mathcal{O}\left(10^{10^9}\right)$ terms in the sum. Evaluating every term in this sum would therefore be impossible with any amount of computing power.[16] Almost all of these terms are zero (involving some "invalid" operator $\langle\alpha_i|O|\alpha_{i+1}\rangle = 0$), and of the nonzero terms, our Monte Carlo procedure will sample only a tiny fraction. It is a miracle that this procedure works at all. Despite the tremendous number of overlooked states, this error in the results of this procedure is small, well-defined, and controlled.

[14] Introducing $M - n$ identity operators means that for each term in Eq. (5.23) there are now $\binom{M}{M-n}$ terms in Eq. (5.25).

[15] As a rough definition, you can think of the weight as the unnormalized probability.

[16] There are, after all, only about 10^{80} atoms in the universe.

5.3.2 Sampling Procedure

Now that we have set up mathematical formulation and configuration space of the SSE QMC method, the next step is to understand the update scheme known as operator-loop updates. These are difficult to describe in the abstract, so we will save much of the explanation to the next few sections: in Sect. 5.4 we apply SSE to the Heisenberg antiferromagnet; in Sect. 5.5 the J-Q_2 model; in Sect. 5.6 the Heisenberg model in an external field; finally, in Sect. 5.7, we combine the work of the previous sections to build an SSE scheme for the J-Q model in an external magnetic field—the J-Q-h model. Nonetheless, it is worthwhile to get a lay of the land before we launch into the details. The configuration in SSE is an initial spin state corresponding to α_0 (in our case using the S^z basis) and a string of operators acting on that state that specify the time-propagated spin states $|\alpha_i\rangle$. Our goal is to sample the space of these operator and spin configurations subject to detailed balance. At all times the configuration must remain "valid," which we define as a configuration where (1) all matrix elements in the time series $\langle\alpha_i|H|\alpha_{i+1}\rangle \neq 0$, and (2) there are periodic boundary conditions in imaginary time (the action of all the operators returns the spin state to $|\alpha_0\rangle$). In SSE the Monte Carlo updates are divided into two separate steps: diagonal updates and off-diagonal updates. Diagonal updates insert and remove diagonal operators (thus sampling the order of the expansion, n) and off-diagonal updates change the operators types and spin state (leaving n fixed).

Diagonal Updates We will loop over each "timeslice"[17] of the configuration from α_0 to α_{M-1}. If no operator is present a slice, we will attempt to insert one. Else if there is a diagonal operator present at the timeslice, then we will attempt to remove it. Else there is an off-diagonal operator present, then it cannot be removed[18] and we will simply update the spin configuration to get the $|\alpha_{i+1}\rangle$ for the next timeslice. The diagonal updates at each timeslice are independent and therefore we can simply loop over the timeslices from $i = 0$ to $M - 1$.

Off-Diagonal Updates Off-diagonal updates change operators from diagonal to off-diagonal and vice versa. The most common way of doing this involves using "operator loops" [1–3, 15], which correspond to cluster updates. These loops can be independent and deterministic, where the structure of the loops are completely determined by the configuration and the loops can be built and flipped independently (as is the case in the Heisenberg model) [1, 3]. In other cases the loops are branching, with the ability to overlap, meaning that loops cannot be built and flipped independently; branching loops must be built sequentially and the choice of how

[17] Here we use "timeslice" to refer to the individual time-propagated states α_i. In the literature this term sometimes refers instead to well-defined intervals of imaginary time composed of many time-propagated states.

[18] Removing an off-diagonal operator (and replacing it with the identity) would result in a zero-valued matrix element and therefore a zero-weighted, "invalid," configuration.

to build them (which branches to choose) can be done with different algorithms including the heat bath algorithm and the directed loop algorithm [2] (the latter is superior and will be used here).

Program Operation and Measurements Although the formalism of quantum Monte Carlo is more complicated than classical Monte Carlo, in the end the simulation itself is a classical Monte Carlo simulation on some extended classical ensemble. The operation of a QMC program with regard to autocorrelations, equilibration, measurement, binning, and estimation of error bars is much the same as a classical Monte Carlo program (as outlined in Sect. 5.2.4). By contrast, construction of the estimators in QMC is somewhat subtle. Since the estimators are intertwined with the diagonal and off-diagonal updates themselves, I will save a detailed discussion of estimators for the end of the first concrete example: the Heisenberg model. This discussion can be found in Sect. 5.4.3.

5.4 The Heisenberg Model

We will now describe how to apply the SSE QMC method to the simplest case—the $S = 1/2$ Heisenberg model. Here I will focus on developing the "theory" of the algorithm in using notation consistent with later sections and address the implementation only briefly. An excellent description of the implementation of SSE for the Heisenberg model complete with pseudocode can be found in Sec. 5.2 of [3] and full FORTRAN implementation of SSE for an $S = \frac{1}{2}$ Heisenberg chain written by Anders Sandvik is available at the address in the footnote.[19]

The Heisenberg model is given by:

$$H = J \sum_{\langle i,j \rangle} \vec{S}_i \cdot \vec{S}_j, \tag{5.28}$$

where $\langle i, j \rangle$ indicates a sum over nearest neighbor spins. We will consider the antiferromagnetic (AFM) case $J > 0$ and restrict ourselves to *bipartite lattices* (such as the square lattice) where the system can be decomposed into two interspersed sublattices with all interactions taking place between sites on opposite sublattices.[20] This is an interacting many-body problem with a highly nontrivial ground state. In $d \geq 2$, the ground state has long-range Néel order (an alternating arrangement of up and down spins), but the Néel state is *not* an eigenstate of the Hamiltonian. Instead, the ground state has Néel correlations along with substantial *quantum fluctuations* which persist even at $T = 0$. This is a marked contrast to the

[19]A full FORTRAN implementation of the SSE method for the $S = \frac{1}{2}$ Heisenberg model can be found here: http://physics.bu.edu/~sandvik/vietri/index.html.

[20]The Heisenberg antiferromagnet will be frustrated and suffer from the sign problem on non-bipartite lattices such as the triangular lattice.

Heisenberg ferromagnet, where the fully polarized ground state *is* an eigenstate of the Hamiltonian. In fact, the Heisenberg AFM is interesting precisely because of these strong quantum fluctuations. Like many interacting many-body Hamiltonians, the Heisenberg model is a sum of local pairwise interactions $\langle i, j\rangle$. All nonzero matrix elements of H will involve just one of these pairs. From here on we will think in terms of these pairs (which we will call *bonds*).

We can expand the total spin operators in terms of their vector components

$$\vec{S}_i \cdot \vec{S}_j = S_i^x S_j^x + S_i^y S_j^y + S_i^z S_j^z. \tag{5.29}$$

We will operate in the S^z basis, so we rewrite S_i^x and S_i^y in terms of the ladder operators $S_i^{\pm}$:

$$\vec{S}_i \cdot \vec{S}_j = S_i^z S_j^z + \frac{1}{2}\left(S_i^+ S_j^- + S_i^- S_j^+\right) \tag{5.30}$$

and then rewrite the Hamiltonian in the S^z basis:

$$H = J \sum_{\langle i,j\rangle} \left[S_i^z S_j^z + \frac{1}{2}\left(S_i^+ S_j^- + S_i^- S_j^+\right)\right]. \tag{5.31}$$

Now we can think of how this Hamiltonian acts on the pairs of spins (bonds) in the S^z basis. Each matrix element will have four *legs*: corresponding to the "before" and "after" spins: $\langle S_{i,f}^z S_{j,f}^z | H | S_{i,0}^z S_{j,0}^z\rangle$. There are $2^4 = 16$ possible combinations of the legs of such an operator, but we can immediately rule out those that do not conserve total S^z spin (such as $\langle + - |H| + +\rangle$) and we are left with just a handful of nonzero matrix element types, two diagonal:

$$\langle \pm \pm |H| \pm \pm\rangle = \frac{J}{4} \tag{5.32a}$$

$$\langle \pm \mp |H| \pm \mp\rangle = -\frac{J}{4} \tag{5.32b}$$

and one off-diagonal:

$$\langle \pm \mp |H| \mp \pm\rangle = \frac{J}{2} \tag{5.32c}$$

Hereafter we will refer to these local matrix elements as *operators* or *vertices*.

For SSE to work, all matrix elements of the Hamiltonian must be ≤ 0, so that the weights of each configuration—Eq. (5.27)—are guaranteed to be positive or zero. To accomplish this, we subtract a constant from the Hamiltonian:

$$H' = H - \frac{J n_b}{4} = J \sum_{\langle i,j \rangle} -\frac{1}{4} + S_i^z S_j^z + \frac{1}{2}\left(S_i^+ S_j^- + S_i^- S_j^+\right) \tag{5.33}$$

This constant can be added back in at the end to produce the correct energy, and will not effect any other quantity. Next, we rotate one of the sublattices by 180° about the $\hat{z}$ axis. This has the effect of rotating $S_i^x \rightarrow -S_i^x$ and $S_i^y \rightarrow -S_i^y$ for all $i \in A$.

$$H'' = J \sum_{\langle i,j \rangle} S_i^z S_j^z - \frac{1}{4} - \frac{1}{2}\left(S_i^+ S_j^- + S_i^- S_j^+\right) \tag{5.34}$$

This sublattice rotation reverses the sign of the off-diagonal terms, but does not alter the spectrum. This works because the interactions are local and pairwise and the lattice is bipartite so each interaction pair includes exactly one site from sublattice A and one site from sublattice B. With these modifications, the matrix elements of the Hamiltonian are now

$$\langle \pm \pm | H | \pm \pm \rangle = 0 \tag{5.35a}$$

$$\langle \pm \mp | H | \mp \pm \rangle = -\frac{J}{2} \tag{5.35b}$$

$$\langle \pm \mp | H | \pm \mp \rangle = -\frac{J}{2} \tag{5.35c}$$

The added constant sets all matrix elements involving parallel spins to zero and the sublattice rotation makes the off-diagonal elements negative. Now that all the nonzero matrix elements are negative, the preconditions for SSE are satisfied.

There are only two kinds of nonzero operators: diagonal and off-diagonal. We can break the Hamiltonian up into its diagonal and off-diagonal components:

$$H = J \sum_b H_{1,b} + H_{2,b} \tag{5.36a}$$

$$H_{1,b} = -\frac{1}{4} + S_{i(b)}^z S_{j(b)}^z \tag{5.36b}$$

$$H_{2,b} = -\frac{1}{2}\left(S_{i(b)}^+ S_{j(b)}^- + S_{i(b)}^- S_{j(b)}^+\right) \tag{5.36c}$$

and write the sum over b, and index of *bonds*. In our program, there will be an array `bonds[2]`[n_b] with n_b rows, each containing $i(b)$ and $j(b)$ to accomplish the sum over $\langle i, j \rangle$. This approach has the advantage that the lattice or dimension can be changed by making a new bond list and no other alterations to the code are required.

By adding the constant we have stumbled upon the definition of the *singlet projection operator*, which can be written

$$P_{i,j} \equiv \frac{1}{4} - \vec{S}_i \cdot \vec{S}_j. \tag{5.37}$$

This operator acts on two spins and projects out the singlet state,

$$P_{i,j} \frac{1}{\sqrt{2}} \left(|\uparrow\downarrow\rangle - |\downarrow\uparrow\rangle \right) = \frac{1}{\sqrt{2}} \left(|\uparrow\downarrow\rangle - |\downarrow\uparrow\rangle \right)$$

which is an eigenstate of $P_{i,j}$. The singlet projection operator is zero acting on parallel spins:

$$P_{i,j} |\uparrow\uparrow\rangle = 0.$$

We can write the modified version of the Heisenberg model from Eq. (5.34) in terms of singlet projection operators:

$$H = -J \sum_{\langle i,j \rangle} P_{i,j} \tag{5.38}$$

Products of the singlet projection operator can be used to construct a wide variety of sign-problem-free, SSE-ready Hamiltonians such as the J-Q family of models [18].

5.4.1 Diagonal Updates

Now we can write Eq. (5.26) in terms of the Heisenberg Hamiltonian:

$$Z = \sum_n \sum_{\{\alpha\}} \frac{(-\beta)^n (M-n)!}{M!} \langle \alpha_0 | O_1 | \alpha_1 \rangle \langle \alpha_1 | O_2 | \alpha_2 \rangle \ldots \langle \alpha_{M-1} | O_M | \alpha_0 \rangle \tag{5.39}$$

where O_p is either $H_{1,b(p)}$, $H_{2,b(p)}$, or the identity. The weight of a given configuration, Eq. (5.27), is then:

$$W(M, n_1, n_2) = \frac{(-\beta)^n (M-n)!}{M!} \left(-\frac{J}{2} \right)^{n_1} \left(-\frac{J}{2} \right)^{n_2}$$

where n_1, n_2 are the number of operators $H_{1,b}$ and $H_{2,b}$, respectively, and $n \equiv n_1 + n_2$. Since both operators have the same value, only the total number of operators contributes to the weight:

$$W(M, n) = \left(\frac{\beta J}{2} \right)^n \frac{(M-n)!}{M!} \tag{5.40}$$

Notice that the sign from the $(-\beta)^n$ has cancelled the sign from the matrix elements, so all weights are guaranteed to be positive and there is no sign problem.

Before we continue, it is worth briefly describing how the spin-operator configurations $\{\alpha\}$ are stored in the SSE QMC program. We start with an initial spin state $|\alpha_0\rangle$ which is stored in an integer array `spins[N]`. We store the operator string in a $2 \times M$ array `operators[2][M]` containing the list of operators $O_0 \ldots O_{M-1}$. Each row of `operators[2][M]` corresponds to a "timeslice" $i = [0, M-1]$, and can contain either the identity (usually encoded as a -1 in both columns) or an operator. In the case of an operator at timeslice i, first column, `operators[0][i]`, contains the operator type and the second column, `operators[1][i]`, contains the index of the bond $b = [0, N_b - 1]$ that the operator acts on. Since this construction is based on a trace, there are periodic boundary conditions in the imaginary time direction: the operators must act to return $|\alpha_0\rangle$ to the initial state (see Fig. 5.1). The operator string completely describes all the changes to the spin state, so an SSE configuration can be specified by storing just the operator string, `operators[2][M]`, and the initial spin state, `spins[N]` ($|\alpha_0\rangle$). Any time-propagated states can be generated on-demand by acting on $|\alpha_0\rangle$ with the operator string. The operator string is a integer array with $2M$ elements and the initial state can be stored as an L^d integer array, other memory requirements are negligible. SSE requires little memory, rarely if ever more than 1 GB; memory use is therefore rarely an issue on modern machines.

For the diagonal update, we loop over each timeslice. If no operator is present, we select a bond at random and attempt to insert an operator. If the spins are parallel, inserting an operator would generate a zero-weight configuration, so we give up and move onto the next timeslice. If the spins are antiparallel, we attempt to insert a diagonal operator. To calculate the probability of accepting this change, we use the Metropolis Algorithm [Eq. (5.13)]

$$A(x \to x') = \min\left(1, \frac{W(x')g(x' \to x)}{W(x)g(x \to x')}\right)$$

First, we must determine the weight of the current state:

$$W_{\text{current}} = \left(\frac{\beta J}{2}\right)^n \frac{(M-n)!}{M!} \tag{5.41}$$

and the weight of the new state (with an extra operator):

$$W_{\text{new}} = \left(\frac{\beta J}{2}\right)^{n+1} \frac{(M-n-1)!}{M!} \tag{5.42}$$

The probability of proposing adding an operator at this particular bond given the configuration is:

$$g_{\text{insert}} = \frac{1}{N_b} \tag{5.43a}$$

since there are N_b bonds and we selected one at random. The probability of proposing to remove it (the reverse process) is simply

$$g_{\text{remove}} = 1 \tag{5.43b}$$

since we will always attempt to remove a diagonal operator. Plug all these into Eq. (5.13) and we find the acceptance probability for inserting an operator

$$A_{\text{insert}} = \min\left(1, \frac{\beta J N_b}{2(M-n)}\right). \tag{5.44}$$

If there is already a diagonal operator present at this timeslice, we will remove with probability A_{Remove}. We can derive this probability starting from

$$W_{\text{new}} = \left(\frac{\beta J}{2}\right)^{n-1} \frac{(M-n+1)!}{M!} \tag{5.45a}$$

and

$$g_{\text{insert}} = \frac{1}{N_b} \tag{5.45b}$$

$$g_{\text{remove}} = 1 \tag{5.45c}$$

Thus:

$$A_{\text{remove}} = \min\left(1, \frac{2(M-n+1)}{\beta J N_b}\right) \tag{5.46}$$

Finally, if there is an off-diagonal operator present at timeslice i, we cannot remove it, since that would generate an invalid configuration. Instead, we flip the pair of spins in $|\alpha_i\rangle$ accordingly to generate $|\alpha_{i+1}\rangle$ and move on to the next timeslice.

5.4.2 Off-Diagonal Updates

The diagonal updates change the order of the series expansion, n. The off-diagonal updates change the spin configurations and operator types by transforming diagonal operators to off-diagonal operators and vice versa. To accomplish this, we build "loops" that traverse the spin configuration. Flipping a loop corresponds to flipping every spin that the loop touches which will also change the operator types along the loop. These features will apply to most operator-loop formulations, but the details from here on will be specific to the Heisenberg model. Changing operator types does not change the total number of operators, so it does not change the weight of the distribution [see Eq. (5.40)]. The probability of proposing flipping a loop is $\frac{1}{4M}$ and the probability of the reverse process is the same. If we were to use the Metropolis Algorithm for deciding to flip loops we would flip every loop with probability $A = 1$ [see Eq. (5.13)]. Flipping every loop would satisfy detailed balance, but it would be extremely inefficient since the configuration is left largely unaltered (all the operator types would remain the same, just the spins would change). Fortunately, the Metropolis Algorithm is just one solution to the detailed balance condition. We can multiply all acceptance rates by any factor u and maintain detailed balance (this will obviously satisfy Eq. (5.12)). Therefore, we will multiply all $A = 1$ by $u = \frac{1}{2}$ and flip each loop with probability one half.[21]

Two examples of SSE configurations for an 8-site $S = \frac{1}{2}$ chain are depicted in Fig. 5.1, which was borrowed from [3]. The horizontal direction represents physical space, with the sites labeled $i = 1 \ldots 8$. The imaginary time direction is vertical (with the time values marked by the leftmost column p). The $|\alpha_0\rangle$ state is at the bottom (with up spins depicted as filled dots and down spins depicted as empty dots). The intermediate spin states, $|\alpha_i\rangle$, are not fully drawn; instead solid lines connect legs on the same site. Since there are no intervening operators, the spin will be the same at all times along a black line. Diagonal vertices are depicted as unfilled rectangles and off-diagonal operators are depicted as solid black rectangles. In Fig. 5.1, the expansion cutoff is $M = 12$ and there are 8 nonidentity operators present ($n = 8$).

We begin a loop by choosing an operator leg as the starting point of our loop and placing an open end of the loop on that leg (this end of the loop will be stationary). We then place the other open end of our loop (the moving end) on the same vertex, on the leg immediately to the right or left of the starting position; this is the "exit" leg (this type of move will be referred to as "switch-and-reverse" in Sect. 5.6). From there, we move the open end of the loop to a leg on a new vertex that is connected to our exit leg by a black line (i.e., a leg on a vertex that acts on the same site, but either earlier or later along the time axis). We then perform another "switch and reverse" to find the exit leg from the second vertex, and repeat this process until the open ends of the loop meet, closing the loop.

In the left panel of Fig. 5.1, the "before" picture, a completed loop is highlighted in red. In the right panel, the "after" picture, the loop (shown as a dashed black line) has been flipped. Here we can see that all the spins on the legs along the loop have been flipped and the operators are flipped once for every time they are visited by the loop. For example, the top left operator has changed from diagonal $\frac{+-}{+-}$ to off-diagonal $\frac{+-}{-+}$. The "middle operator" is still diagonal; it has been visited twice by

[21]One might ask why we would use $u = \frac{1}{2}$ and not some other fraction. It is easy to convince oneself that this is optimal. Multiplying by 0 or 1 would clearly generate bad updates and it seems logical that there should be symmetry between u and $1-u$, thus the optimal choice would be where $u = 1 - u$ therefore $u = \frac{1}{2}$.

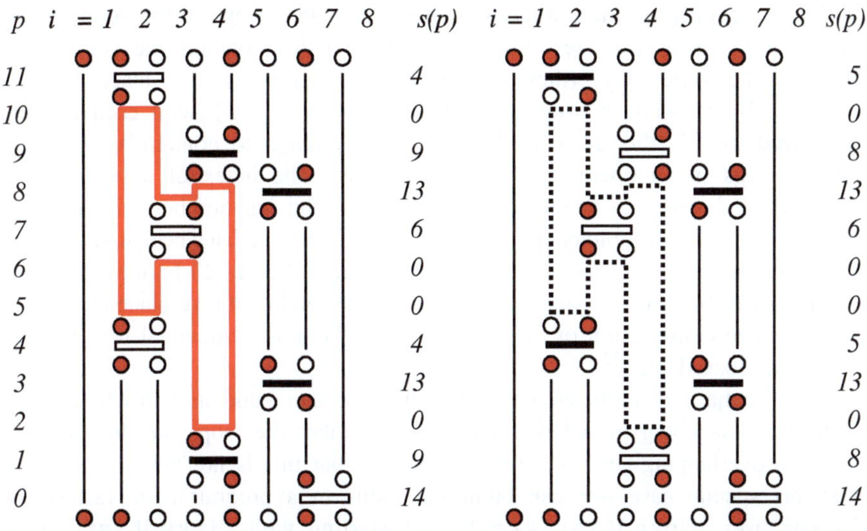

Fig. 5.1 An example SSE configuration for an 8-site chain (left) with a loop before it has been flipped and (right) after. Filled circles represent spin-up sites and open circles represent spin-down sites; rectangle outlines represent diagonal operators and solid rectangles represent off-diagonal operators. This figure appeared as Fig. 61 in Sec. 5.2.2 of [3] (reprinted under fair use)

the loop (the spins on every leg have been flipped), therefore it has changed from diagonal with spins $\frac{-+}{-+}$ to off-diagonal and back to diagonal, but now with flipped spins $\frac{+-}{+-}$.

A few comments here about the loops for the Heisenberg model. Each leg is a member of exactly one loop and a loop will never visit the same leg twice. As a corollary, to this statement, it does not matter where on the loop you start, or which direction you move, you will always construct the same loop. Loops must close in order to return the system to a valid configuration. Loops can update the initial spin state $|\alpha_0\rangle$; this will occur when the loop wraps around the periodic boundary conditions in the imaginary time direction "passing through" the $\tau = 0$ state $|\alpha_0\rangle$. Such a loop is not depicted in Fig. 5.1, but one could be constructed for that configuration by starting from the up spin on the "bottom" of the operator at $p = 4$. Loops can update the total spin of the $|\alpha_0\rangle$ configuration if it wraps around (passing through $\tau = 0$) an odd number of times. The evolution along the imaginary time direction represents real quantum mechanical dynamics of the system where all conservations laws are obeyed. I emphasize again that by contrast, simulation time dynamics (i.e., the consecutive SSE configurations generate by the loop updates) do *not* follow Hamiltonian dynamics and conservations laws—like total S^z conservation—are *not* obeyed.

Implementation The details of how this method is implemented as a computer program are described in detail and with pseudocode in Sec. 5.2.3 of [3]. I will include only a brief description here. The first step of the loop update is to build a data structure (called a linked list) that contains the connections between the vertex legs (these connections correspond to the black lines in Fig. 5.1). Begin with an array `linkedList[4*cutoff]` with each element initialized to -2 (this array is therefore large enough to contain an entry for all four legs of an operator at every timeslice). The index of a given leg is $i = 4p + l$ where p is the timeslice and l is the leg number with the legs numbered $\frac{2\,3}{0\,1}$. The completed list will have $4n$ elements that have been updated such that `linkedList[i] = linkedList[linkedList[i]]`; the remaining elements will be "empty," since the identity operators do not take place in the loop updates. The construction of the linked list will be almost identical for the other models discussed later.[22]

One we have constructed the linked list, we can build and flip loops. The isotropic Heisenberg model is a special case where the loops are deterministic and non-overlapping (in Sect. 5.6 we will see that this is not always the case). For this special case, we can build the loops in a programmatic fashion by looping over `linkedList[4*cutoff]` starting with `linkedList[0]`. If `linkedList[i]=-2`, then it is not connected to anything and we can move on $i' = i + 1$. When we come across `linkedList[i]`≥ 0, we then build a loop. Before we build the loop, we flip a coin to decide whether to flip the loop. Starting with the leg i, we follow the linked list around the loop until we reach the starting point and the loop is closed. As we follow the loop, we mark those legs as visited (by setting `linkedList[j]`$= -1$) and update the spin and operator type.

Once we are finished with the loops, we can do one more thing. Spins that are not connected to any vertex like the spin at position $i = 1$ in Fig. 5.1 do not affect the energy. At the end of the loop update, we can flip these unconnected spins with probability half (the reason for using $A = \frac{1}{2}$ here is the same as the reason for flipping the loops with probability $\frac{1}{2}$).[23]

5.4.3 Observables in SSE

Now that I have described a concrete example of SSE method, I can describe how to extract expectation values from the Markov chain of SSE configurations. The remarks in this section are quite general and will apply to both the Heisenberg model and the other applications of SSE discussed in Sects. 5.5 to 5.7. Any observable that is diagonal in the basis can be extracted relatively easily and inexpensively. All the simulations I will describe here use the S^z basis, so observables such as the S^z

[22]In some of my simulations I have also stored the spin configuration of the operator legs in an array `legs[4*cutoff]`. This imposes a cost in memory use and is not strictly necessary, but is nonetheless useful for debugging when the operator types become more complicated.

[23]This step is not strictly necessary, but it helps.

magnetization and S^z spin correlations can be extracted easily by performing an unweighted average over the initial configurations, $|\alpha_0\rangle$:

$$\langle O \rangle = \langle \alpha_0 | O | \alpha_0 \rangle \tag{5.47}$$

In this and all the following equations for diagonal estimators the bracketed term on the left-hand side is the expectation value and the right-hand side of the equation describes how to make a measurement on a single SSE configuration. For increased statistics, one can average over the full SSE configuration including the time-propagated states $\langle \alpha_{i \neq 0} \rangle$ [3, Sec. 5.2.4]:

$$\langle O \rangle = \frac{1}{M} \sum_{p=0}^{M-1} \langle \alpha_p | O | \alpha_p \rangle \,. \tag{5.48}$$

This extracts more information from each SSE configuration, but also involves more computational effort, and measurements performed on the time-propagated states are highly correlated. The choice of whether or not to average over time-propagated states is a variation of the "measurement frequency" problem we discussed in Sect. 5.2.4. For simple and cheap observables such as the magnetization this improved estimator is almost always worthwhile, but for more complicated quantities like correlation functions it can be expensive and can even grow to dominate the computational cost of the simulation.

Off-diagonal observables do not have an obvious classical analog. Constructing estimators for off-diagonal observables is therefore somewhat tricky. For this reason, I have mostly avoided measuring off-diagonal observables. In many cases diagonal observables suffice or there are easier-to-measure substitutes for an off-diagonal observable. For example, in the isotropic zero-field Heisenberg model, there is full $O(3)$ rotational symmetry, so therefore $\langle |\vec{S}|^2 \rangle = 3 \langle |S_z|^2 \rangle$ and there is no need to directly calculate the X and Y components (which are off-diagonal). Alternatively, in the Heisenberg model with a field, we cannot rely on symmetry to extract the spin correlation length in the XY plane, but the spin stiffness, ρ_s, is easy to calculate and provides a way of detecting long-range or quasi-long-range spin correlations (see Sect. 4.4.1) [19, 20]. Estimators for two-point off-diagonal observables can in some circumstances be constructed by following the motion of the open ends of the loops [21] and for more complicated off-diagonal observables using complicated and expensive auxiliary loops. Off-diagonal observables that appear as terms in the Hamiltonian, however, can often be measured in a simple way. The best example is the energy; H is obviously not diagonal in the S^z basis (or the whole problem would be trivial), but there is a simple formula for $\langle H \rangle$. First we start from the general formula for determining any observable O from the density matrix [22, p. 189]:

$$\langle O\rangle = \frac{1}{Z}Tr\,\{\rho O\} = \frac{1}{Z}\sum_{\alpha_0}\langle\alpha_0|\rho O|\alpha_0\rangle \tag{5.49}$$

Now we can plug in $O = H$, Taylor expand as we did for the partition function in Eq. (5.21) and then rewrite the sum in terms of the new index $n' = n + 1$

$$\langle E\rangle = \frac{1}{Z}\sum_{\alpha_0}\sum_{n=0}^{\infty}\frac{(-\beta)^n}{n!}\langle\alpha_0|(H)^{n+1}|\alpha_0\rangle$$

$$\langle E\rangle = \frac{1}{Z}\sum_{\alpha_0}\sum_{n'=1}^{\infty}\frac{(-\beta)^{n'-1}}{(n'-1)!}\langle\alpha_0|(H)^{n'}|\alpha_0\rangle = \frac{1}{Z}\sum_{\alpha_0}\sum_{n'=1}^{\infty}\frac{n'}{-\beta}\frac{(-\beta)^{n'}}{n'!}\langle\alpha_0|(H)^{n'}|\alpha_0\rangle$$

We can now extend this sum to include $n' = 0$ (the $n' = 0$ term vanishes) to recover a sum that has the same terms as the stochastic series expansion itself [Eq. (5.21)],

$$\langle E\rangle = \frac{1}{Z}\sum_{n=0}^{\infty}\left(-\frac{n}{\beta}\right)\frac{(-\beta)^n}{n!}\langle\alpha_0|(H)^n|\alpha_0\rangle\,, \tag{5.50}$$

but with the value of each term multiplied by a factor of $-n/\beta$. Thus $\langle E\rangle$ is simply the expectation value of the order of the expansion, n (i.e., the total number of operators) [3, Sec. 5.1.3]:

$$\langle E\rangle = -\frac{\langle n\rangle}{\beta} \tag{5.51}$$

We can follow a similar procedure to find the squared energy, $\langle E^2\rangle$:

$$\langle E^2\rangle = \frac{1}{Z}\sum_{\alpha_0}\sum_{n=0}^{\infty}\frac{(-\beta)^n}{n!}\langle\alpha_0|(H)^{n+2}|\alpha_0\rangle$$

$$\langle E^2\rangle = \frac{1}{Z}\sum_{\alpha_0}\sum_{n'=2}^{\infty}\frac{(-\beta)^{n'-2}}{(n'-2)!}\langle\alpha_0|(H)^{n'}|\alpha_0\rangle$$

$$\langle E^2\rangle = \frac{1}{Z}\sum_{\alpha_0}\sum_{n'=2}^{\infty}\frac{n(n-1)}{\beta^2}\frac{(-\beta)^{n'}}{n'!}\langle\alpha_0|(H)^{n'}|\alpha_0\rangle$$

$$\langle E^2\rangle = \frac{1}{Z}\sum_{n=0}^{\infty}\frac{n(n-1)}{\beta^2}\frac{(-\beta)^n}{n!}\langle\alpha_0|(H)^n|\alpha_0\rangle \tag{5.52}$$

And therefore $\langle E^2\rangle$ is simply [3, Sec. 5.1.3]

$$\langle E^2 \rangle = -\frac{\langle n(n-1) \rangle}{\beta} \tag{5.53}$$

Note: the estimator for $\langle E^2 \rangle$ is not simply the square of the $\langle E \rangle$ estimator due to its derivation from the series expansion. Using $\langle E^2 \rangle$ and $\langle E \rangle$ we can then calculate the specific heat using the energy fluctuations formula [3, Sec. 5.1.3]:

$$C_v = \beta^2 \left(\langle E^2 \rangle - \langle E \rangle^2 \right) = \langle n^2 \rangle - \langle n \rangle^2 - \langle n \rangle \tag{5.54}$$

Other examples of off-diagonal observables with simple estimators in SSE include the VBS bond susceptibility [23] and the *spin stiffness*, ρ_s [19, 20]. The nature of the spin stiffness and a discussion of how it is calculated in SSE simulations can be found in Sect. 4.4.1.

5.5 The J-Q_2 Model

The J-Q_2 model is composed of a Heisenberg-like J term and a four-spin Q term made up of the product of two singlet projection operators.[24]

$$H_{JQ} = -J \sum_{\langle i,j \rangle} P_{i,i+1} - Q \sum_{\langle i,j,k,l \rangle} P_{i,j} P_{k,l} \tag{5.55}$$

Here $\langle i, j \rangle$ represents a sum over nearest neighbors (as before), and $\langle i, j, k, l \rangle$ represents a sum over sites in a line (in 1D) or on a plaquette ($\begin{smallmatrix} k & l \\ i & j \end{smallmatrix}$ and $\begin{smallmatrix} j & l \\ i & k \end{smallmatrix}$ in 2D). This Hamiltonian is a numerical method in its own right. Prior to its development [24], there were no known sign-problem-free Hamiltonians with a deconfined quantum critical point. The J-Q model is a "designer Hamiltonian" which is interesting not because it is a physically realistic model of a real material, but because it provides a QMC-friendly environment in which to study the valence-bond solid (VBS) state and the phase transition from the Néel to the VBS state. The J-Q_x models are part of a family [18] of sign-problem free Hamiltonians[25] can be built from products of singlet projection operators [see Eq. (5.37)].

The J term in the Hamiltonian is identical to the Heisenberg model, Eq. (5.38). We can break this portion up into pieces in the same manner as we did for the pure Heisenberg model:

$$H = J \sum_b H_{1,b} + H_{2,b} \tag{5.56a}$$

[24] There is also a variant that uses three singlet projection operators called the J-Q_3 model.

[25] Another word for sign-problem-free is Marshall positive.

$$H_{1,b} = -\frac{1}{4} + S^z_{i(b)} S^z_{j(b)} \tag{5.56b}$$

$$H_{2,b} = -\frac{1}{2}\left(S^+_{i(b)} S^-_{j(b)} + S^-_{i(b)} S^+_{j(b)}\right) \tag{5.56c}$$

For the Q term, we have two pairwise interactions that can also be broken up into diagonal $D_{ij} \equiv \frac{1}{4} - S^z_i S^z_j$ and off-diagonal $O_{ij} \equiv \frac{1}{2}\left(S^+_i S^-_j + S^-_i S^+_j\right)$ pieces:

$$H_Q = -Q\sum_i (D_{i,j} + O_{i,j})(D_{k,l} + O_{k,l})$$

$$H_Q = -Q\sum_i (D_{i,j} D_{k,l} + D_{i,j} O_{k,l} + O_{i,j} D_{k,l} + O_{i,j} O_{k,l})$$

So we can break H_Q up into 4 "operators" which can be (independently) act diagonally or off-diagonally on the first and second pairs of spins:

$$H_{3,b} = -Q\left(\frac{1}{4} - S^z_i S^z_j\right)\left(\frac{1}{4} - S^z_k S^z_l\right) \quad \text{(DD)} \tag{5.57a}$$

$$H_{4,b} = -\frac{Q}{2}\left(\frac{1}{4} - S^z_i S^z_j\right)(S^+_k S^-_l + S^-_k S^+_l) \quad \text{(DO)} \tag{5.57b}$$

$$H_{5,b} = -\frac{Q}{2}\left(S^+_i S^-_j + S^-_i S^+_j\right)\left(\frac{1}{4} - S^z_k S^z_l\right) \quad \text{(OD)} \tag{5.57c}$$

$$H_{6,b} = -\frac{Q}{4}\left(S^+_i S^-_j + S^-_i S^+_j\right)(S^+_k S^-_l + S^-_k S^+_l) \quad \text{(OO)} \tag{5.57d}$$

This looks complicated but breaks down into relatively simple rules for Q interactions:

1. Bonds where either pair of spins are parallel $S^z_i = S^z_j$ or $S^z_k = S^z_l$ have zero weight. (Note that there is no restriction on the relationship between S^z_j and S^z_k).
2. Operators acting on between S^z_i and S^z_j can be diagonal or off-diagonal.
3. Operators acting on S^z_k and S^z_l can be diagonal or off-diagonal.
4. All nonzero Q-type operators have the same energy (and weight): $-\frac{Q}{4}$.

We can write all the possible matrix elements in terms of the S^z spins on each of the eight legs, for a total of $2^8 = 256$ possible matrix elements. As before, we can throw away all those that violate S^z conservation as having zero weight, have a further restriction: the singlet projection operator is only nonzero when $S^z_i = -S^z_j$, each pair of legs corresponding to i, j and k, l must be antiparallel. In the end there are only four types of 4-spin operators, each with the same weight, $Q/4$. They can be diagonal or off-diagonal in the first pair (i, j) and (independently) diagonal or off-diagonal in the second pair (k, l). Thus the four types will be abbreviated DD, OD, DO, and OO.

Now we are ready to write the partition function of a given configuration from Eq. (5.25).

$$Z = \sum_{n}\sum_{\{\alpha\}} \frac{(-\beta)^{n_J+n_Q}(M-n_J-n_Q)!}{M!} \times \langle\alpha_0|\,O_1\,|\alpha_1\rangle\,\langle\alpha_1|\,O_2\,|\alpha_2\rangle \ldots \langle\alpha_{M-1}|\,O_M\,|\alpha_0\rangle \tag{5.58}$$

where O_p can be the identity or any $H_{x,b}$ that connects $\langle\alpha_{p-1}|$ to $|\alpha_p\rangle$. The weight of a configuration depends on n_J and n_Q, the total numbers of J-type and Q-type operators present:

$$W(M, n_J, n_Q) = \frac{(M-n_J-n_Q)!}{M!}\left(\frac{\beta J}{2}\right)^{n_J}\left(\frac{\beta Q}{4}\right)^{n_Q} \tag{5.59}$$

5.5.1 Diagonal Updates

For the diagonal update, we will proceed in an almost identical manner to the pure Heisenberg case, so here we will highlight the changes. For a diagonal update, we go through each time step and check if there is an operator there. If no operator is present, we will attempt to insert a diagonal operator. However, now there are two different types of diagonal operators J-type [Eq. (5.56b)] and Q-type [Eq. (5.57a)]. We will flip a coin to decide which type of operator to attempt to insert.[26]

If we decide to attempt to insert $H_{1,b}$, we first check that the spins for bond b are antiparallel $S^z_{i(b)} = -S^z_{j(b)}$. If they are antiparallel, we can proceed in calculating the acceptance probability using the Metropolis Algorithm, Eq. (5.13):

$$W(M, n_J+1, n_Q) = \frac{(M-n_J-n_Q-1)!}{M!}\left(\frac{\beta J}{2}\right)^{n_J+1}\left(\frac{\beta Q}{4}\right)^{n_Q} \tag{5.60a}$$

$$g_{\text{insert-J}} = \frac{1}{2N_b} \tag{5.60b}$$

$$g_{\text{remove-J}} = 1 \tag{5.60c}$$

$$A_{\text{insert-J}} = \min\left(1, \frac{\beta J N_b}{M-n_J-n_Q}\right) \tag{5.60d}$$

[26]It might seem strange or inefficient to decide this by chance without using any information about the state (like if a Q-type operator even can be inserted), but this method of deciding is simple, unbiased, and (most importantly) makes it easy to calculate the proposal probability g.

Note that the proposal probability, $g_{\text{insert-J}}$ has an extra factor of $1/2$ compared to the Heisenberg case; this comes from the fact that we flipped a coin to decide between inserting a J-type and Q-type operator.

If we decide to attempt to insert a Q-type operator, we check that both pairs of spins are antiparallel, i.e., $S^z_{i(b)} = -S^z_{j(b)}$ and $S^z_{k(b)} = -S^z_{l(b)}$ If so, we can calculate the probability of accepting this change:

$$W(M, n_J, n_Q + 1) = \frac{(M - n_J - n_Q - 1)!}{M!} \left(\frac{\beta J}{2}\right)^{n_J} \left(\frac{\beta Q}{4}\right)^{n_Q+1} \tag{5.61a}$$

$$g_{\text{insert-Q}} = \frac{1}{2N_b} \tag{5.61b}$$

$$g_{\text{remove-Q}} = 1 \tag{5.61c}$$

$$A_{\text{insert-Q}} = \min\left(1, \frac{\beta Q N_b}{2(M - n_J - n_Q)}\right) \tag{5.61d}$$

If there is already a diagonal operator present at a timeslice, we will attempt to remove it with probability

$$W(M, n_J - 1, n_q) = \frac{(M - n_J - n_Q + 1)!}{M!} \left(\frac{\beta J}{2}\right)^{n_J-1} \left(\frac{\beta Q}{4}\right)^{n_Q} \tag{5.62a}$$

$$g_{\text{insert-J}} = \frac{1}{2N_b} \tag{5.62b}$$

$$g_{\text{remove-J}} = 1 \tag{5.62c}$$

$$A_{\text{remove-J}} = \min\left(1, \frac{M - n_J - n_Q + 1}{\beta J N_b}\right) \tag{5.62d}$$

for J-type operators. For Q-type operators the probability is:

$$W(M, n_J, n_Q - 1) = \frac{(M - n_J - n_Q + 1)!}{M!} \left(\frac{\beta J}{2}\right)^{n_J} \left(\frac{\beta Q}{4}\right)^{n_Q-1} \tag{5.63a}$$

$$g_{\text{insert-Q}} = \frac{1}{2N_b} \tag{5.63b}$$

$$g_{\text{remove-Q}} = 1 \tag{5.63c}$$

$$A_{\text{remove-Q}} = \min\left(1, \frac{2(M - n_J - n_Q + 1)}{\beta Q N_b}\right) \tag{5.63d}$$

Note that a Q-type operator can only be removed if it is completely diagonal in both the first and second pair (type 3) otherwise and removing it will generate an invalid (zero-weight) configuration, like $\langle \uparrow\downarrow\uparrow\downarrow\ |\mathbb{I}|\ \uparrow\downarrow\downarrow\uparrow \rangle$.

5.5.2 *Off-Diagonal Updates*

For the J-Q model, the off-diagonal (loop) updates proceed in an identical manner to the pure Heisenberg model. The data structures are all the same, except now allowing for 8 legs per vertex. When a loop encounters a Q-type operator, it flips the "half" it encounters from diagonal to off-diagonal or vice versa. For example, let us say a loop enters the following DO Q-type operator.

$$\begin{array}{cc} +- & +- \\ \hline +- & -+ \\ & \uparrow \end{array} \Rightarrow \begin{array}{cc} +- & +- \\ \hline +- & +- \\ & \uparrow\downarrow \end{array} \tag{5.64}$$

It does a "switch and reverse" and changes it to a DD operator just as it would if the right half of this vertex had been a Heisenberg J-type operator.

5.6 The Heisenberg Model in an External Field

We will now formulate an SSE QMC procedure for the (sublattice-rotated) Heisenberg model [Eq. (5.34)] in the presence of an external magnetic field in the z direction given by:

$$H = -J \sum_{<i,j>} P_{i,j} - h \sum_i S_i^z \tag{5.65}$$

where h represents the external magnetic field and $J > 0$ represents a nearest neighbor antiferromagnetic coupling constant. Surprisingly, this will require more modifications to our SSE method than the J-Q_2 model.

We want to solve this model using SSE, so we must again meet the constraint that all matrix elements must be negative or zero. To get the Hamiltonian into SSE shape, we follow a similar procedure to the pure Heisenberg model, but with some important differences. This approach follows closely the approach for the anisotropic Heisenberg model in an external field discussed in [2].

To start we will substitute in for $P_{i,j}$ (the sublattice-rotated version).[27]

$$H_{Jh} = -J \sum_{\langle i,j \rangle} \left[\frac{1}{4} - S_i^z S_j^z + \frac{1}{2} \left(S_i^+ S_j^- + S_i^- S_j^+ \right) \right] - h \sum_i S_i^z \tag{5.66}$$

[27] In the sublattice-rotated version of $P_{i,j}$, the S^z operators have the opposite sign from the ladder operators.

We will again think in terms of n_b pairs of spins (bonds),[28] so we will replace S_i^z with $(S_i^z + S_j^z)/2$ and merge the two sums into a sum over bonds labeled x

$$H_{Jh} = \sum_x \Big[-J\left(\frac{1}{4} - S^z_{i(x)}S^z_{j(x)} + \frac{1}{2}\left(S^+_{i(x)}S^-_{j(x)} + S^-_{i(x)}S^+_{j(x)}\right)\right) \\ -h_b\left(S^z_{i(x)} + S^z_{j(x)}\right)\Big] \tag{5.67}$$

defining $h_b \equiv \frac{h}{2d}$ to avoid double counting (since each spin now appears in $2d$ bonds). Now we can break H up into its diagonal and off-diagonal components:

$$H = \sum_x \left(H_{1,x} + H_{2,x}\right) \tag{5.68a}$$

$$H_{1,x} = -J\left(\frac{1}{4} - S^z_{i(x)}S^z_{j(x)}\right) - h_b\left(S^z_{i(x)} + S^z_{j(x)}\right) \tag{5.68b}$$

$$H_{2,x} = -\frac{1}{2}\left(S^+_{i(x)}S^-_{j(x)} + S^-_{i(x)}S^+_{j(x)}\right) \tag{5.68c}$$

To ensure that all nonzero matrix elements will be negative, we will add a constant, $-C$, to the diagonal term:

$$H' = -Cn_b + \sum_x \left(H_{1,x} + H_{2,x}\right) \tag{5.69}$$

$$H'_{1,x} = \sum_x \left[-C - J\left(\frac{1}{4} - S^z_{i(x)}S^z_{j(x)}\right) - h_b\left(S^z_{i(x)} + S^z_{j(x)}\right)\right]. \tag{5.70}$$

In order to get the correct energy we can add an offset of n_bC at the end.

Now that we have broken H up, let us write all the nonzero local matrix elements. There are four different types of nonzero configurations: three diagonal

$$\langle ++|H_1|++\rangle = -C - h_b \tag{5.71a}$$

$$\langle --|H_1|--\rangle = -C + h_b \tag{5.71b}$$

$$\langle \pm\mp|H_1|\pm\mp\rangle = -C - \frac{J}{2}, \tag{5.71c}$$

and one off-diagonal:

$$\langle \pm\mp|H_2|\mp\pm\rangle = -\frac{J}{2}. \tag{5.71d}$$

[28] In 1D the number of bonds n_b is just N, in 2D it is $2N$, etc.

Now assume $h_b \geq 0$ and solve for C to guarantee that all these matrix elements are negative.

$$\begin{aligned}
-C + h_b &\leq 0 \\
C &\geq h_b \\
C_0 &= h_b \\
C &= C_0 + \epsilon
\end{aligned} \tag{5.72}$$

Here we have introduced a constant ϵ that represents the excess over the minimum value of C, $C_0 = h_b$.

Now we can write down the weights for all the operators (matrix elements):

$$W_1 = \langle + + |H_1| + + \rangle = -\epsilon - 2h_b \tag{5.73a}$$

$$W_{-1} = \langle - - |H_1| - - \rangle = -\epsilon \tag{5.73b}$$

$$W_3 = \langle \pm \mp |H_1| \pm \mp \rangle = -\frac{J}{2} - h_b - \epsilon \tag{5.73c}$$

$$W_4 = \langle \pm \mp |H_2| \mp \pm \rangle = -\frac{J}{2} \tag{5.73d}$$

Here the indices are chosen with care toward implementation; all odd-numbered operators are diagonal, and W_1 is the spin-inverted version of W_{-1}, the same for W_3 and W_4. The matrix elements (or operators) are more complicated than for the zero-field case [see Eq. (5.35)]; here we have two additional nonzero matrix elements and the weights for the matrix elements are all different.

5.6.1 Diagonal Updates

The diagonal update procedure will be broadly similar to the pure Heisenberg model with a few key differences. We can calculate the weight of a given configuration based on the number of operators it contains of each type,

$$\begin{aligned}
W(n_{-1}, n_1, n_3, n_4) = {} & \frac{(-\beta)^n (M-n)!}{M!} (-\epsilon)^{n_{-1}} (-\epsilon - 2h_b)^{n_1} \\
& \times \left(-\epsilon - \frac{J}{2} + h_b\right)^{n_3} \left(-\frac{J}{2}\right)^{n_4}
\end{aligned}$$

where M is the "cutoff" and $n = n_1 + n_2 + n_3 + n_4$. The minus signs from the individual operator weights and the $(-\beta)^n$ prefactor cancel:

$$W(n_{-1}, n_1, n_3, n_4) = \frac{\beta^n (M-n)!}{M!} (\epsilon)^{n_{-1}} (\epsilon + 2h_b)^{n_1} \left(\epsilon + \frac{J}{2} + h_b\right)^{n_3} \left(\frac{J}{2}\right)^{n_4} \tag{5.74}$$

Unlike in the pure Heisenberg model, there are now three different types of diagonal operators. At each timeslice if no operator is present, we select a bond at random and attempt to insert the appropriate type of operator (-1, 1 or 3) and accept the change with a probability that obeys the detailed balance condition. If the spins are antiparallel, we will attempt to insert a type 3 operator: $\langle \uparrow\downarrow |H_3| \uparrow\downarrow\rangle = -\frac{J}{2} - \epsilon - h_b$. The weight of the configuration with an extra type 3 operator is:

$$W(n_{-1}, n_1, n_3+1, n_4) = \frac{\beta^{n+1} (M-n-1)!}{M!} (\epsilon)^{n_{-1}} (\epsilon + 2h_b)^{n_1} \times \left(\epsilon + \frac{J}{2} + h_b\right)^{n_3+1} \left(\frac{J}{2}\right)^{n_4}$$

The probability of proposing this change is still just $1/n_b$. The fact that there are multiple types of operators does not, in this case, affect the probability of proposing this change. The spin configuration decides what operator type we attempt to insert, so once this bond has been selected, the probability of proposing this specific operator is one. Using the Metropolis Algorithm, the acceptance probability for inserting this operator:

$$A^{(3)}_{\text{insert}} = \min\left[1, \frac{2n_b\beta}{M-n}\left(\epsilon + \frac{J}{2} + h_b\right)\right] \tag{5.75}$$

If the spins are both up ($|\uparrow\uparrow\rangle$), then we try to insert a type 1 operator at this location: $\langle \uparrow\uparrow |H_1| \uparrow\uparrow\rangle = -\epsilon - 2h_b$. The weight of the proposed new configuration is

$$W(n_{-1}, n_1+1, n_3, n_4) = \frac{\beta^{n+1} (M-n-1)!}{M!} (\epsilon)^{n_{-1}} (\epsilon + 2h_b)^{n_1+1} \times \left(\epsilon + \frac{J}{2} + h_b\right)^{n_3} \left(\frac{J}{2}\right)^{n_4}$$

The acceptance probability is then

$$A^{(1)}_{\text{insert}} = \min\left(1, \frac{2n_b\beta}{M-n}(\epsilon + 2h_b)\right) \tag{5.76}$$

If the spins are both down ($|\downarrow\downarrow\rangle$), we will attempt to insert a type -1 operator: $\langle \downarrow\downarrow |H_{-1}| \downarrow\downarrow\rangle = -\epsilon$. The weight for the new configuration is:

$$W(n_{-1}+1, n_1, n_3, n_4) = \frac{\beta^{n+1}(M-n-1)!}{M!} (\epsilon)^{n_{-1}+1} (\epsilon + 2h_b)^{n_1} \times \left(\epsilon + \frac{J}{2} + h_b\right)^{n_3} \left(\frac{J}{2}\right)^{n_4}$$

The acceptance probability is therefore

$$A^{(-1)}_{\text{insert}} = \min\left(1, \frac{2n_b\beta\epsilon}{M-n}\right) \tag{5.77}$$

If there is already an operator present and that operator is diagonal (of type -1, $+1$, or 3), then we will remove it with probability:

$$A^{(-1)}_{\text{remove}} = \min\left(1, \frac{M-n+1}{2n_b\beta}\frac{1}{\epsilon}\right) \tag{5.78a}$$

$$A^{(1)}_{\text{remove}} = \min\left(1, \frac{M-n+1}{2n_b\beta}\frac{1}{\epsilon+2h_b}\right) \tag{5.78b}$$

$$A^{(3)}_{\text{remove}} = \min\left(1, \frac{M-n+1}{2n_b\beta}\frac{1}{\epsilon+\frac{J}{2}+h_b}\right) \tag{5.78c}$$

Finally, if a type 4 (off-diagonal) operator is present, we simply flip the spins and move on to the next timeslice (this part is identical to the pure Heisenberg model).

5.6.2 Off-Diagonal Updates

For the off-diagonal updates we need an entirely new procedure. In the pure Heisenberg model we could build loops and flip each of them with probability $1/2$ because diagonal and off-diagonal operators had the same weight, so changing any number of diagonal operators to off-diagonal operators (and vice versa) did not alter the weight of the configuration. With the magnetic field, this is no longer true. We could still construct the loops as we did before, and decide how to flip them using the Metropolis condition. The problem is that loops change many operators and therefore change the energy by a large amount causing the loop acceptance probability to become extremely small. Fortunately, there is another paradigm for constructing loops.

In the loop updates for pure Heisenberg model the loops are constructed *deterministically* and the loops are then flipped *stochastically*. For the J-h model the loop will instead be *constructed stochastically*. When a loop is closed, detailed balance is already satisfied and the loops can be flipped with 100% probability. This procedure has the advantage that no time is wasted constructing loops that are never flipped.

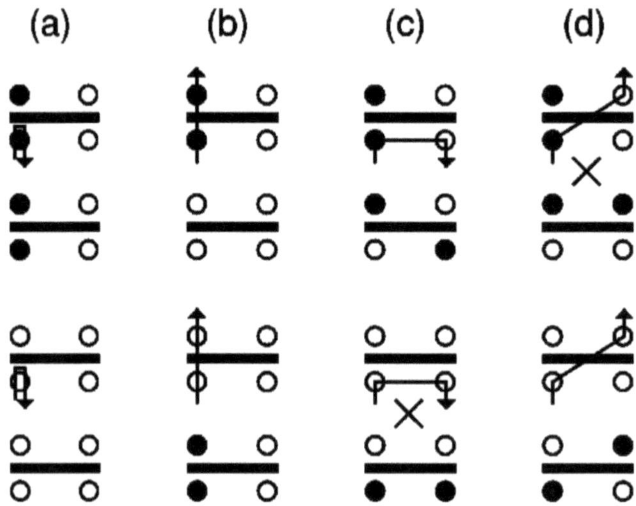

Fig. 5.2 A representation of the possible exit paths when the entrance is on the lower left leg. Column (**a**) depicts a *bounce*, exiting on the lower left leg; column (**b**) depicts *continue straight*, exiting on the upper left leg; column (**c**) depicts *switch and reverse*, exiting on the lower right leg; and column (**d**) depicts *switch and continue*, exiting on the upper right leg. The case of a diagonal operator $\langle\uparrow\downarrow|H|\uparrow\downarrow\rangle$ is shown on the first two rows with the initial configuration followed by the final configuration. The last two rows show the same processes for the all-down spin configuration $\langle\downarrow\downarrow|H|\downarrow\downarrow\rangle$. Places where the resulting operator has zero weight are marked with an X. This figure appeared as Fig. 3 in [2] (reprinted under fair use)

How can do we actually construct the loops stochastically? The basic idea works like this. You start by choose a vertex leg (a leg of an operator) at random and flip that entrance leg. For example, let us say you start (enter) on the lower left leg of a vertex $\frac{+-}{+-}$. The resulting vertex, $\frac{+-}{--}$, is invalid (i.e., has zero weight). To correct this we must choose a second leg to flip (the "exit" leg). The four ways to do this are depicted in the first two rows of Fig. 5.2: we could (a) "bounce"—exit on the leg we came in on, producing the same vertex we started with, $\frac{+-}{+-}$; (b) "continue straight," producing $\frac{--}{--}$; (c) "switch and reverse" exiting on the lower right leg and producing $\frac{+-}{-+}$; or (d) "switch and continue," exiting on the upper right leg and producing $\frac{++}{--}$ (which is an invalid operator). We choose one of these exit legs in manner that obeys detailed balance. Then we follow the exit leg to a new entrance leg on the next connected vertex (as is done in the conventional operator-loop updates) and repeat this process until we return to the leg we started on. It may happen that the same vertex is visited more than once. Figure 5.2 was borrowed from the original directed loop paper by Syljuåsen and Sandvik [2].

By contrast, in the Heisenberg model, only the "switch-and-reverse" option (Fig. 5.2c) is ever used and the loops are therefore deterministic and non-overlapping. Loops in the pure Heisenberg model have the additional property that they are *independent*, which here means that flipping a loop does not affect paths of subsequent loops or the likelihood of flipping them. In the pure Heisenberg case we could build all the loops at once and decide whether to flip them later, now we must build and flip the loops one after another.

Heat Bath Updates

The more difficult question is how to choose the exit leg in a manner that satisfies detailed balance. The simplest approach that satisfies detailed balance uses the so-called *heat bath* equations.[29] In the heat bath approach the probability of exiting on leg i is given by the weight of the operator produced by exiting on that leg, W_i, normalized by the sum of the weights of operators produced by various exit legs.

$$P_i = \frac{W_i}{\sum W_j} \tag{5.79}$$

This solution is simple but it is suboptimal [2] because it results in a high proportion of bounces: exiting on the leg you came in on. Bounces waste time: you build a loop, but spend a substantial amount of time retracing your steps rebuilding the same portions of the loop.

The heat bath equations are just one of infinite solutions to the detailed balance condition. Mathematically and physically, all these solutions are equally correct, but practically, some of them will produce a better Markov chain than others (as measured by autocorrelation times). A trivial example of a bad solution to detailed balance is $P(x \rightarrow x') = 0$ for all $x \neq x'$ (all zero acceptance probabilities). This choice satisfies the detailed balance condition, Eq. (5.9), but clearly does not result in a successful Markov chain.

Directed Loop Updates

To arrive at any particular solution to the detailed balance condition we must impose additional constraints; and if we are doing that, why not choose constraints that maximize the usefulness of the solution? If we decide to minimize the probability of backtracking (which we will call the *bounce* probability) we arrive at the *directed loop* algorithm [2]. Although it has not been proven rigorously, it makes intuitive sense that minimizing backtracking will likely lead to the most efficient updates. In [2] it is shown that for the anisotropic Heisenberg model in an external field, directed loop updates are more efficient than heat bath updates, and reproduce Heisenberg-like deterministic loops in the isotropic zero-field limit.

---

[29] The heat bath solution to detailed balance is a good example of a solution to the detailed balance condition that is not the Metropolis Algorithm.

The following derivation follows [2]. The are two basic principles for setting up the directed loops equations: (1) the weight of each vertex should equal the sum of the weights of the ways of exiting that vertex from a single entrance leg and (2) the weights of time-reversed processes should be equal. We can start with a vertex we will call vertex 1 with weight W_1, and enter on a leg, which we will call leg 1. The weight of vertex 1 should add up to the weights of all the possible exits.

$$W_1 = a_{11} + a_{12} + a_{13} + a_{14} \tag{5.80}$$

Here, the coefficient a_{ij} is the weight entering the vertex on leg i and exiting on leg j. Exiting vertex 1 on leg 2 produces a different vertex, which we will call vertex 2 and has weight W_2. The weight of vertex 2 must add up to the weights of all the possible exits when entering from leg 2.

$$W_2 = a_{21} + a_{22} + a_{23} + a_{24} \tag{5.81}$$

We can continue this for all vertices that can be produced by entering vertex 1 on leg 1, creating a 4×4 matrix equation:

$$\begin{pmatrix} a_{11} & a_{12} & a_{13} & a_{14} \\ a_{21} & a_{22} & a_{23} & a_{24} \\ a_{31} & a_{32} & a_{33} & a_{34} \\ a_{41} & a_{42} & a_{43} & a_{44} \end{pmatrix} \begin{pmatrix} 1 \\ 1 \\ 1 \\ 1 \end{pmatrix} = \begin{pmatrix} W_1 \\ W_2 \\ W_3 \\ W_4 \end{pmatrix}$$

We can simplify by setting $a_{ij} = a_{ji}$ using time-reversal symmetry and replacing the bounce probabilities $a_{ii} = b_i$. Yielding a real symmetric matrix equation that enforces detailed balance:

$$\begin{pmatrix} b_1 & a_{12} & a_{13} & a_{14} \\ a_{12} & b_2 & a_{23} & a_{24} \\ a_{13} & a_{23} & b_3 & a_{34} \\ a_{14} & a_{24} & a_{34} & b_4 \end{pmatrix} \begin{pmatrix} 1 \\ 1 \\ 1 \\ 1 \end{pmatrix} = \begin{pmatrix} W_1 \\ W_2 \\ W_3 \\ W_4 \end{pmatrix} \tag{5.82}$$

The vertex weights W_i are known. If the starting vertex is W_1, and we have entered on leg 1, the exit probabilities are

$$P_1 = \frac{b_1}{W_1} \tag{5.83a}$$

$$P_2 = \frac{a_{12}}{W_1} \tag{5.83b}$$

$$P_3 = \frac{a_{13}}{W_1} \tag{5.83c}$$

$$P_4 = \frac{a_{14}}{W_1} \tag{5.83d}$$

Now we need only to find a solution for the a_{ij} that satisfies this equation with the additional constraint that $a_{ij}, b_i \geq 0$. There are 10 unknowns and only 4 equations, so there are multiple (infinite) solutions, a large subset of which will include only nonnegative coefficients [2]. As discussed before, we will attempt to find solutions that minimize the bounce probabilities b_i. Generically these equations would be solved either by hand or computer program and hardcoded into the QMC procedure. In this case the solution to the directed loop equations for the anisotropic Heisenberg model in an external field has been solved in [2], and we can use that solution here. In the isotropic case ($\Delta = 1$) the problem can be reduced to two sets of directed loop equations. Since one of the exit legs always produces an invalid vertex (see Fig. 5.2), each directed loop matrix equation can be further reduced to a 3×3 matrix equation. The first set of resulting coupled equations can be written[30]:

$$W_4 = b_1 + a + b \tag{5.84a}$$

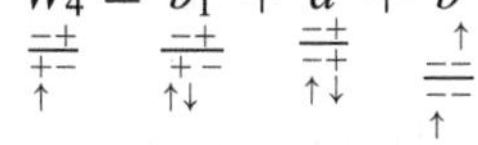

$$W_3 = a + b_2 + c \tag{5.84b}$$

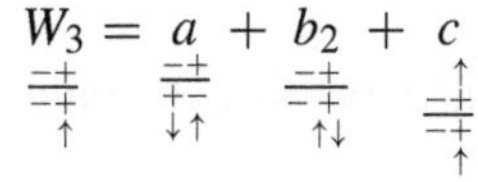

$$W_{-1} = b + c + b_3 \tag{5.84c}$$

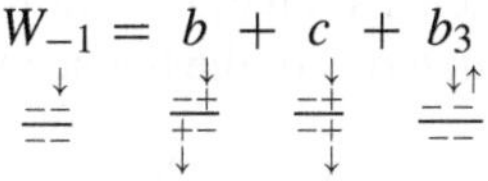

where the vertex weights are labeled in accordance with the numbering scheme in Eq. (5.73). The vertex diagram under each W_i indicates the vertex type. The vertex symbol under each exit weight denotes the type of vertex made by exiting on that leg, with the entrance and exit legs indicated by inward and outward facing arrows (respectively). The second set of equations[31] can be written:

$$W_4 = b_1' + a' + b' \tag{5.85a}$$

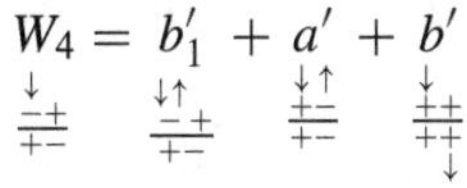

$$W_3 = a' + b_2' + c' \tag{5.85b}$$

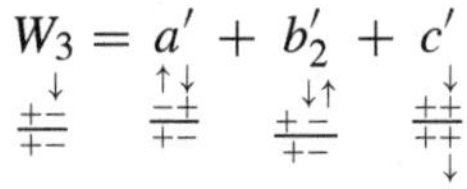

$$W_1 = b' + c' + b_3' \tag{5.85c}$$

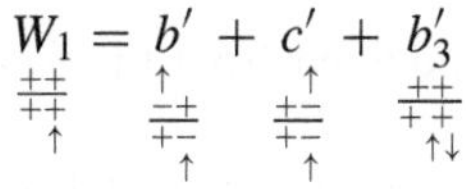

[30]This first set corresponds to the upper left quadrant of Fig. 8 of [2].

[31]This second set corresponds to the lower left quadrant of Fig. 8 of [2].

These relations will hold under time-reversal symmetry (swapping the top and bottom of the vertices $\frac{-+}{+-} \rightarrow \frac{+-}{-+}$) and mirror symmetry (swapping the left and right of the vertices $\frac{+-}{+-} \rightarrow \frac{-+}{-+}$). Using those symmetries any combination of initial vertex and entrance leg can be described in terms of these diagrams.

Before we actually solve for the weights, let us consider an example. Say we come across a type 3 vertex that looks like this $\frac{+-}{+-}$, entering from the top left. By time-reversal symmetry and mirror symmetry this is equivalent to the scenario described by Eq. (5.84b). There are three valid exits: bounce, switch-and-reverse, and continue straight, with the following weights:

$$P_{\text{switch-and-reverse}} = \frac{a}{W_3} \tag{5.86a}$$

$$P_{\text{bounce}} = \frac{b_2}{W_3} \tag{5.86b}$$

$$P_{\text{continue-straight}} = \frac{c}{W_3} \tag{5.86c}$$

Switch-and-continue is forbidden because it would generate a zero-weighted operator: $\frac{--}{++}$.

Now to solve for the coefficients. First, plug in for the weights using Eq. (5.73) and solve for the coefficients in terms of the bounce weights:

$$a = \frac{1}{2} + \frac{h_b}{2} + \frac{-b_1 - b_2 + b_3}{2} \tag{5.87a}$$

$$b = -\frac{h_b}{2} + \frac{-b_1 + b_2 - b_3}{2} \tag{5.87b}$$

$$c = \frac{h_b}{2} + \epsilon + \frac{b_1 - b_2 - b_3}{2} \tag{5.87c}$$

$$a' = \frac{1}{2} - \frac{h_b}{2} + \frac{-b'_1 - b'_2 + b'_3}{2} \tag{5.87d}$$

$$b' = \frac{h_b}{2} + \frac{-b'_1 + b'_2 - b'_3}{2} \tag{5.87e}$$

$$c' = \frac{3h_b}{2} + \epsilon + \frac{b'_1 - b'_2 - b'_3}{2} \tag{5.87f}$$

This form has been slightly modified from the form in [2] in order to allow for $J = 0$, which we will use in Sect. 5.7.

Now we rely on the solutions in Tab. I of [2]. Their solution breaks up the parameter space into six regions depending on the values of Δ, J, and h. For the isotropic case ($\Delta = 1$), we are only concerned with Region III and Region IV. The solution for Region III, where $\Delta = 1$ and $0 \leq h_b \leq J$ is given by:

$$a = \frac{J}{2} \tag{5.88a}$$

$$b = 0 \tag{5.88b}$$

$$c = \epsilon \tag{5.88c}$$

$$a' = \frac{J}{2} - \frac{h_b}{2} \tag{5.88d}$$

$$b' = \frac{h_b}{2} \tag{5.88e}$$

$$c' = \frac{3h_b}{2} + \epsilon \tag{5.88f}$$

with the bounce probabilities:

$$b_1 = 0 \tag{5.88g}$$

$$b_2 = h_b \tag{5.88h}$$

$$b_3 = 0 \tag{5.88i}$$

$$b_1' = 0 \tag{5.88j}$$

$$b_2' = 0 \tag{5.88k}$$

$$b_3' = 0 \tag{5.88l}$$

Note that (unlike the heat balance solution) almost all bounce probabilities are zero and we even recover the deterministic Heisenberg update scheme in the limit $h \to 0$, with bounce probabilities vanishing. The solution for region IV, where $\Delta = 1$ and $h_b > J$, has a, b, c unchanged:

$$a = \frac{J}{2} \tag{5.89a}$$

$$b = 0 \tag{5.89b}$$

$$c = \epsilon \tag{5.89c}$$

$$a' = 0 \tag{5.89d}$$

$$b' = \frac{J}{2} \tag{5.89e}$$

$$c' = h_b + \frac{J}{2} + \epsilon \tag{5.89f}$$

with the bounce probabilities:

$$b_1 = 0 \tag{5.89g}$$

$$b_2 = h_b \tag{5.89h}$$

$$b_3 = 0 \tag{5.89i}$$

$$b'_1 = 0 \tag{5.89j}$$

$$b'_2 = 0 \tag{5.89k}$$

$$b'_3 = h_b - J \tag{5.89l}$$

Once again most of the bounce probabilities are zero, but there are always some nonzero bounce weights.

5.7 The *J*-*Q*-*h* Model

We will now combine the work from the previous sections to construct an SSE simulation scheme for the J-Q_2 model with an external field—the J-Q-h model, which is given by:

$$H = -J \sum_{<i,j>} P_{i,j} - Q \sum_{<i,j,k,l>} P_{i,j} P_{k,l} - h \sum_i S_i^z. \tag{5.90}$$

where $P_{i,j}$ is a singlet projection operator, $\langle i, j \rangle$ refers to a sum over all nearest neighbor pairs, and $\langle i, j, k, l \rangle$ refers to $j = i + 1$, $k = i + 2$, and $l = i + 3$ in 1D and horizontal $\substack{k\ l \\ i\ j}$ and vertical $\substack{j\ l \\ i\ k}$ bonds on a plaquette in 2D. As before, we will use the sublattice-rotated version of $P_{i,j}$. This section will be quite brief; all the ingredients for the SSE procedure for the J-Q-h model have been developed in the preceding sections and now we need only to combine them.

To get this Hamiltonian in SSE-shape, we will separate it into two broad categories of operators: 2-spin operators (incorporating the J and h terms) and 4-spin operators (involving the Q term). The 2-spin operators will be treated as they were in the Heisenberg model with an external field and the 4-spin operators will be treated as they were in the zero-field J-Q_2 case. A more useful form of the Hamiltonian can thus be written in terms of the Hamiltonian for the J-Q model [Eq. (5.55)] with the Hamiltonian for the Heisenberg model with an external field [Eq. (5.67)]:

$$H_{JQh} = \sum_{\langle i,j \rangle} \left[P_{i,j} - h_b \left(S_i^z + S_j^z \right) \right] - Q \sum_{\langle i,j,k,l \rangle} P_{i,j} P_{k,l} \tag{5.91}$$

where once again $h_b \equiv \frac{h}{2d}$ to avoid double counting. We can now make a list of all the two-spin [Eq. (5.73)] and four-spin [Eq. (5.57)] operators:

$$W_1 = \langle + + | H | + + \rangle = \epsilon + 2h_b \qquad \text{(D)} \tag{5.92a}$$

$$W_{-1} = \langle - - |H| - - \rangle = \epsilon \qquad \text{(D)} \qquad (5.92b)$$

$$W_3 = \langle \pm \mp |H| \pm \mp \rangle = \frac{J}{2} + h_b + \epsilon \qquad \text{(D)} \qquad (5.92c)$$

$$W_4 = \langle \pm \mp |H| \mp \pm \rangle = \frac{J}{2} \qquad \text{(O)} \qquad (5.92d)$$

$$W_5 = \langle \pm \mp \pm \mp |H| \pm \mp \pm \mp \rangle = \frac{Q}{4} \qquad \text{(DD)} \qquad (5.92e)$$

$$W_6 = \langle \pm \mp \pm \mp |H| \mp \pm \pm \mp \rangle = \frac{Q}{4} \qquad \text{(OD)} \qquad (5.92f)$$

$$W_7 = \langle \pm \mp \pm \mp |H| \pm \mp \mp \pm \rangle = \frac{Q}{4} \qquad \text{(DO)} \qquad (5.92g)$$

$$W_8 = \langle \pm \mp \pm \mp |H| \mp \pm \mp \pm \rangle = \frac{Q}{4} \qquad \text{(OO)} \qquad (5.92h)$$

Here we have already cancelled the minus sign in all these weights with the minus sign in the prefactor $(-\beta)^n$. The indices are chosen with care toward implementation: all odd-numbered operators act diagonally on the first bond and all even-numbered operators act off-diagonally on the first bond. Here not all of the two-spin operators respect spin-inversion symmetry (because of the h term); W_1 is the spin-inverted version of W_{-1}, the same for W_3 and W_4.

5.7.1 Diagonal Updates

We can use Eq. (5.92) to write an expression for the weight of a configuration:

$$W = \frac{(-\beta)^n (M-n)!}{M!} (\epsilon)^{n_{-1}} (\epsilon + 2h_b)^{n_1} \left(\frac{J}{2} + h_b + \epsilon\right)^{n_3} \left(\frac{J}{2}\right)^{n_4} \left(\frac{Q}{4}\right)^{n_q} \tag{5.93}$$

where $n_q \equiv n_5 + n_6 + n_7 + n_8$ and $n \equiv \sum n_i$. With the operator types and weights established, let us consider how to proceed with the updates. We will begin with the operator string stored in the array `operators[2][m]`: a list with M elements containing n operators and $M - n$ identity operators (marked by a zero). If a nonidentity operator is present at timeslice i, `operators[0][i]` contains the bond number, b, that the operator acts on and `operators[1][i]` contains the type of operator, i.e., $-1, 1, 3, 4, 5, 6, 7, 8$.

In the diagonal update, the program loops over all m rows of `operators`. For each row, if there is no operator present, it will attempt to insert one; If there is a diagonal operator present, it will attempt to remove it, and if there is an off-diagonal operator present, it will update the list of spins accordingly.

Operator Insertion

At each timeslice if there is no operator present, we will attempt to insert an operator with acceptance probability $A(x \to x')$ based on the Metropolis solution to the detailed balance condition in Eq. (5.13). The procedure is identical to the J-Q procedure except we now have the 3 diagonal 2-spin operators from the J-h case. The probability of proposing to insert an operator any particular bond is

$$g_{\text{insert}} = \frac{1}{2n_b} \tag{5.94a}$$

and the probability of proposing to remove it is

$$g_{\text{remove}} = 1 \tag{5.94b}$$

Thus, the acceptance probability when inserting a diagonal operator of type O_p be it J-type or Q-type is

$$A_{\text{insert}}^{(O_p)} = \min\left(1, \frac{2n_b\beta}{M-n} W_{O_p}\right). \tag{5.95}$$

Operator Removal

To remove an operator, we can follow nearly identical procedure to operator insertion, but in reverse. The acceptance probability for removing a diagonal operator of type O_p

$$A_{\text{remove}} = \min\left(1, \frac{M-n+1}{2n_b\beta} \frac{1}{W_{O_p}}\right) \tag{5.96}$$

5.7.2 Directed Loop Updates

Here we will use a combination of deterministic and directed loop updates. When a loop encounters a 2-spin operator, it will choose the exit leg using the solution to the directed loop equations in Sect. 5.6 (see also [2]), when it encounters a 4-spin operator, it will do the efficient switch-and-reverse moves from the zero-field J-Q model.

5.8 Supplementary Procedures

The methods described in this chapter are in general very effective, but in some cases additional "bolt-on" procedures can improve the Monte Carlo sampling. In the case of the *J*-*Q*-*h* model, especially in 1D, we found that the directed loop simulations tended to get "frozen" in magnetization states. Simulations required extremely long times to update magnetization sectors. As a result, the bins were correlated, the autocorrelation time diverged and it was impossible to make accurate estimates of statistical error. At low temperatures, this problem was so severe that a simulation might never leave a magnetization sector once it became stuck. An example of this problem can be seen in Fig. 5.3, a plot of the magnetization density for a *J*-*Q*-*h* chain as a function of *h* that appeared in a preliminary proceedings report [25]. In Fig. 5.3 the magnetization curves appear "rough" in a manner that resembles statistical error, but is in fact the result of this freezing process. To remedy the sticking problem we used two additional techniques, *quantum replica exchange* and *β-doubling*, which we will describe in the following sections.

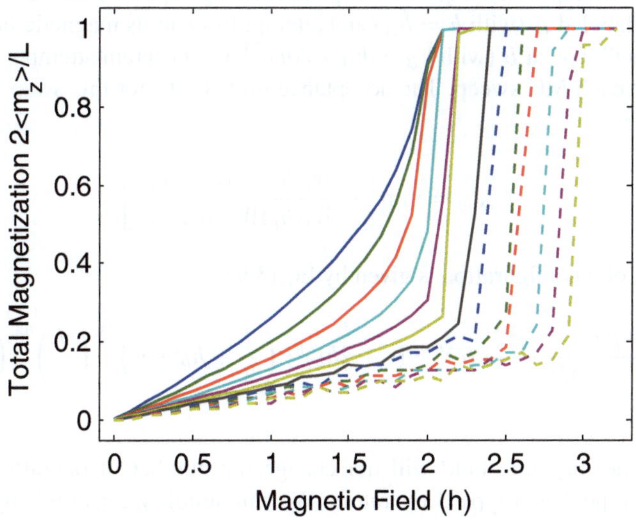

Fig. 5.3 Scaled magnetization plotted as a function of applied magnetic field for a range of different values of $q = Q/J$. From the left (solid blue), $q = 0.0,\ 0.1,\ 0.2 \ldots\ 1.2$. Computed using QMC with $L = 140$ and open boundary conditions. The statistical error for all points is exactly zero; simulations become stuck in magnetization states, causing incorrect estimates of statistical error. This figure originally appeared in [25] as Fig. 3 (reprinted under fair use)

5.8.1 Quantum Replica Exchange

Replica exchange [26], also known as parallel tempering, is a parallelized multicanonical Monte Carlo method. In conventional Monte Carlo, the simulation parameters like temperature are fixed; in replica exchange, N_r simulations (replicas) are run in parallel on a mesh of temperatures. The Monte Carlo updates to these simulations proceed as normal, but after each complete Monte Carlo sweep, the replicas are allowed to swap temperatures in a manner that obeys detailed balance within the extended multicanonical ensemble.[32] Replica exchange is a sort of continuous simulated annealing—simulations which are in a metastable state will "random walk" to a higher temperature where they can become unstuck.

In quantum replica exchange [17] some parameter that is typically fixed (which may or may not be temperature) is sampled stochastically. In this work on the J-Q-h model, the simulations tended to become stuck in magnetization sectors (see Fig. 5.3) and we were already interested in results over a range of magnetic fields, so it was natural to stochastically sample the magnetic field. We begin by initializing N_r simulations, each with a magnetic field $h_i = h_0 + \Delta h \times i$. At the end of each Monte Carlo sweep we do a quantum replica exchange update where select a replica at random, labeled A (with $h = h_A$) and attempt to swap its magnetic field with its "neighboring" replica B (with $h_B = h_A + \Delta h$).[33] The program attempts N_r of these swaps after each MC sweep. The acceptance probability for this swap is given by [17]:

$$A_{\text{swap}} = \min\left[1, \frac{W_i(h_{i+1})W_{i+1}(h_i)}{W_i(h_i)W_{i+1}(h_{i+1})}\right] \tag{5.97}$$

The weight of a configuration is given by Eq. (5.93):

$$W = \frac{(-\beta)^n(M-n)!}{M!}(\epsilon)^{n-1}(\epsilon + 2h_b)^{n_1}\left(\frac{J}{2} + h_b + \epsilon\right)^{n_3}\left(\frac{J}{2}\right)^{n_4}\left(\frac{Q}{4}\right)^{n_q} \tag{5.93}$$

Swapping the magnetic field will not change the number of operators, only the weights of type 1 and type 3 operators. We can simplify Eq. (5.93) by replacing the constant portions with a constant G:

---

[32]This method can also be done just a single replica, sampling the temperature stochastically without swapping, but then typically a bias in the temperature acceptance rates must be imposed in order to ensure that the desired temperature regime is sampled.

[33]In principle, we could allow swaps between any two replicas, but in practice, the acceptance rates of swaps involving large changes in field is nearly zero. Considering swaps only between neighbors results in a higher acceptance rate without violating the detailed balance condition.

$$W = G\,(\epsilon + 2h_b)^{n_1}\left(\frac{J}{2} + \epsilon + h_b\right)^{n_3} \tag{5.98}$$

Now we can calculate the weights of the original configurations:

$$W_A(h_A) = G_A\,(\epsilon + 2h_A)^{n_1^A}\left(\frac{J}{2} + \epsilon + h_A\right)^{n_3^A} \tag{5.99a}$$

$$W_B(h_B) = G_B\,(\epsilon + 2h_B)^{n_1^B}\left(\frac{J}{2} + \epsilon + h_B\right)^{n_3^B} \tag{5.99b}$$

and the field-swapped configurations:

$$W_A(h_B) = G_A\,(\epsilon + 2h_B)^{n_1^A}\left(\frac{J}{2} + \epsilon + h_B\right)^{n_3^A} \tag{5.99c}$$

$$W_B(h_A) = G_B\,(\epsilon + 2h_A)^{n_1^B}\left(\frac{J}{2} + \epsilon + h_A\right)^{n_3^B} \tag{5.99d}$$

Now we can plug these equations into Eq. (5.97) [17]:

$$\begin{aligned} A(h_A, h_B) &= \min\left[1, \frac{W_A(h_B)W_B(h_A)}{W_A(h_A)W_B(h_B)}\right] \\ A(h_A, h_B) &= \min\left[1, \left(\frac{\epsilon + 2h_B}{\epsilon + 2h_A}\right)^{n_1^A - n_1^B}\left(\frac{J + 2\epsilon + 2h_B}{J + 2\epsilon + 2h_A}\right)^{n_3^A - n_3^B}\right] \end{aligned} \tag{5.100}$$

to arrive at an equation for the acceptance rate.

Some care has to be taken to avoid dividing by zero if $h_A = 0$ and $\epsilon = 0$. In general, this situation can be avoided by setting ϵ to some small nonzero value like 0.1. Nonetheless, I wrote my code to account for the possibility that $\epsilon = 0$. To do this, I first set $\epsilon = 0$ in Eq. (5.100)

$$P(h_A, h_B) = \min\left[1, \left(\frac{h_B}{h_A}\right)^{n_1^A - n_1^B}\left(\frac{J + 2h_B}{J + 2h_A}\right)^{n_3^A - n_3^B}\right]$$

When h_A is zero, n_1^A must be equal to zero (because those operators have zero weight).

$$P(h_A = 0, h_B) = \min\left[1, \left(\frac{h_B}{h_A}\right)^{-n_1^B}\left(\frac{J + 2h_B}{J}\right)^{n_3^A - n_3^B}\right]$$

If $n_1^B > 0$, then configuration B cannot accept a zero magnetic field, since that would result in a zero-weighted configuration. If $n_1^B = 0$, then configuration B *can* accept a zero magnetic field since

$$\lim_{h_A \to 0} \left(\frac{h_B}{h_A} \right)^{-n_1^B} = \begin{cases} 0 & n_1^B > 0 \\ 1 & n_1^B = 0 \end{cases} \tag{5.101}$$

Therefore the acceptance probability is

$$P(h_A = 0, h_B) = \begin{cases} 0 & n_1^B > 0 \\ \min\left[1, \left(\frac{J+2h_B}{J}\right)^{n_3^A - n_3^B}\right] & n_1^B = 0 \end{cases} \tag{5.102}$$

Similar considerations must be undertaken to avoid entering a zero-weighted configuration when $J = 0$ and have been implemented in my program. It is worth noting here that in my simulation h_B is always > 0 because $h_B = h_A + \Delta h$.

5.8.2 β Doubling

One of the most difficult aspects of Monte Carlo can be producing the first configurations drawn from the appropriate distribution, a process called equilibration. Typically, when initializing a simulation we start with a completely random guess for the initial state. In the Ising model this would just be a random spin state and in SSE the initial configuration is a randomized spin state and an empty operator string. We then commence Monte Carlo updates and wait until a number of "initialization" or "equilibration" sweeps have been completed before recording data. In most cases this works, but it can be problematic. These initial guesses are drawn from the infinite-temperature distribution of states, and immediately commencing regular Monte Carlo sweeps is equivalent to performing a quench, which can result in getting stuck in metastable, nonequilibrium states.[34] In classical Monte Carlo, this problem can be solved with *simulated annealing* [27], where the temperature is initially high and is gradually lowered over time, which helps ensure that the true equilibrium state is reached.

Simulated annealing can be performed in QMC as well, but instead I have used a related procedure known as β-doubling (where β refers to the inverse temperature) [20]. In β-doubling, the simulation starts at a high temperature, like $\beta_0 = 1$ and it is allowed to equilibrate for some number of Monte Carlo steps. At the end of this equilibration the configuration, $\mathcal{C}$, is not just the spin state, also the extended imaginary time configuration (i.e., the operator string):

[34] Strictly speaking, simulation time is not the same as physical time, but the effect is often similar.

$$\mathcal{C}(\beta_0) = \langle\alpha_0|O_0|\alpha_1\rangle\,\langle\alpha_1|O_1|\alpha_2\rangle\ldots\langle\alpha_{M-1}|O_{M-1}|\alpha_0\rangle \tag{5.103}$$

Next we will double $\beta_1 = 2\beta_0$, but before we continue with the Monte Carlo updates, we can make an improved guess for an equilibrium configuration at β_1. We expect the number of nonidentity operators in the operator string to be roughly $n \propto \beta$, and therefore, a good guess for $\mathcal{C}(\beta_1)$ is an operator string twice as long, i.e., $[\mathcal{C}(\beta_0)]^2$. In practical terms, this means we simply append the operator string array to itself and double the expansion cutoff, M. Getting to β_f is then a matter of repeatedly doubling β and the configuration, with $\beta_{i+1} = 2\beta_i$. This is a simplified version of the procedure used in [20] (where measurements were made at each value of β). This procedure is simple to implement and greatly enhances equilibration, especially in difficult-to-equilibrate systems like the J-Q-h model.

5.9 Pseudorandom Number Generation

Monte Carlo relies on random numbers, but computers do not behave randomly. Instead, we rely on *pseudorandom* number generators which use algorithms to generate an unpredictable sequence of numbers from a "seed" that is used as a starting point. To eliminate the need for an external random number seed, my simulations have been in most cases designed to use the system clock (via the C++ function `time(NULL)`) as a seed. For the quantum replica exchange calculations, the seeds for the parallel processes consist of the product of the system clock and the process ID which will generate a unique seed for each system.

For a random number generator I have used the C++ port of the Mersenne Twister algorithm available here.[35] Although the simpler linear congruential random number generator [6, Sec. 4.3] is usually sufficient for Monte Carlo, for some seeds it can have short periods, whereas the Mersenne Twister [28] has been proven to have a period of $2^{19937} - 1$. In any case, my benchmarks indicated that this professional implementation of the Mersenne Twister is (marginally) faster than my homemade implementation of the linear congruential random number generator.

References

1. A.W. Sandvik, J. Kurkijärvi, Phys. Rev. B **43**, 5950 (1991). http://dx.doi.org/10.1103/PhysRevB.43.5950
2. O.F. Syljuåsen, A.W. Sandvik, Phys. Rev. E **66**, 046701 (2002). http://dx.doi.org/10.1103/PhysRevE.66.046701

[35] C++ implementation of the Mersenne Twister: http://www.bedaux.net/mtrand/.

3. A.W. Sandvik, in *American Institute of Physics Conference Series*, ed. by A. Avella, F. Mancini, vol. 1297, (2010), pp. 135–338, http://arxiv.org/abs/1101.3281. http://dx.doi.org/10.1063/1.3518900
4. P. Weinberg, M. Bukov, SciPost Phys. **2**, 003 (2017). http://dx.doi.org/10.21468/SciPostPhys.2.1.003
5. N. Metropolis, S. Ulam, J. Am. Stat. Assoc. **44**, 335 (1949). http://dx.doi.org/10.1080/01621459.1949.10483310
6. S. Brandt, *Data Analysis: Statistical and Computational Methods for Scientists and Engineers*, 3rd edn. (Springer, 1998).
7. H. Suwa, S. Todo, Phys. Rev. Lett. **105**, 120603 (2010). http://dx.doi.org/10.1103/PhysRevLett.105.120603
8. N. Metropolis, A.W. Rosenbluth, M.N. Rosenbluth, A.H. Teller, E. Teller, J. Chem. Phys. **21**, 1087 (1953). http://dx.doi.org/10.1063/1.1699114
9. K.-H. Barth, Interview of Marshall Rosenbluth, Niels Bohr Library and Archives (American Institute of Physics, 2003). https://www.aip.org/history-programs/niels-bohr-library/oral-histories/28636-1
10. J.E. Gubernatis, Phys. Plasmas **12**, 057303 (2005). http://dx.doi.org/10.1063/1.1887186
11. J.E. Gubernatis, AIP Conf. Proc. **690**, 3 (2003). http://dx.doi.org/10.1063/1.1632111
12. W.K. Hastings, Biometrika, 97 (1970). http://dx.doi.org/10.1093/biomet/57.1.97
13. M.N. Rosenbluth, AIP Conf. Proc. **690**, 22 (2003). http://dx.doi.org/10.1063/1.1632112
14. A.W. Sandvik, "The stochastic series expansion method," Website: http://physics.bu.edu/~sandvik/research/ssehistory.html. Accessed 18 Oct 2017
15. H.G. Evertz, Adv. Phys. **52**, 1 (2003). http://dx.doi.org/10.1080/0001873021000049195
16. A. Iaizzi, K. Damle, A.W. Sandvik, Phys. Rev. B **95**, 174436 (2017). http://dx.doi.org/10.1103/PhysRevB.95.174436
17. P. Sengupta, A.W. Sandvik, D.K. Campbell, Phys. Rev. B **65**, 155113 (2002). http://dx.doi.org/10.1103/PhysRevB.65.155113
18. R.K. Kaul, R.G. Melko, A.W. Sandvik, Annu. Rev. Condens. Matter Phys. **4**, 179 (2013). http://dx.doi.org/10.1146/annurev-conmatphys-030212-184215
19. T. Einarsson, H.J. Schulz, Phys. Rev. B **51**, 6151 (1995). http://dx.doi.org/10.1103/PhysRevB.51.6151
20. A.W. Sandvik, Phys. Rev. B **66**, 024418 (2002). http://dx.doi.org/10.1103/PhysRevB.66.024418
21. A. Dorneich, M. Troyer, Phys. Rev. E **64**, 066701 (2001). http://dx.doi.org/10.1103/PhysRevE.64.066701
22. J. Sakurai, J. Napolitano, *Modern Quantum Mechanics* (Addison-Wesley, 2011). https://books.google.com/books?id=N4I-AQAACAAJ
23. S. Jin, A.W. Sandvik, Phys. Rev. B **87**, 180404 (2013). http://dx.doi.org/10.1103/PhysRevB.87.180404
24. A.W. Sandvik, Phys. Rev. Lett. **98**, 227202 (2007). http://dx.doi.org/10.1103/PhysRevLett.98.227202
25. A. Iaizzi, A.W. Sandvik, J. Phys. Conf. Ser. **640**, 012043 (2015). http://stacks.iop.org/1742-6596/640/i=1/a=012043
26. K. Hukushima, K. Nemoto, J. Phys. Soc. Jpn. **65**, 1604 (1996). http://dx.doi.org/10.1143/JPSJ.65.1604
27. S. Kirkpatrick, C.D. Gelatt, M.P. Vecchi, Science **220**, 671 (1983). http://dx.doi.org/10.1126/science.220.4598.671
28. M. Matsumoto, T. Nishimura, ACM Trans. Model. Comput. Simul. **8**, 3 (1998). http://dx.doi.org/10.1145/272991.272995

Chapter 6
Conclusions

I have presented a comprehensive study of the J-Q model in the presence of an external magnetic field in both one and two dimensions. To accomplish this I have developed a quantum Monte Carlo program based on the stochastic series expansion with directed loop updates and incorporating quantum replica exchange. In the one-dimensional J-Q model (see Chap. 2) there are magnetization jumps to saturation (metamagnetism) above a critical coupling ratio $q_{\min}$ [1, 2]. This is the first reported example of metamagnetism occurring in the absence of frustration or intrinsic anisotropy and is another example (beside the Néel-VBS transition) of a behavior of that usually occurs in frustrated systems made accessible to large-scale quantum Monte Carlo study using the sign-problem-free J-Q model. I show that the magnetization jumps are caused by the onset of attractive interactions between magnons (spin flips on a background of uniformly polarized spins) and derive an exact analytical solution for $q_{\min} = 2/9$ based on a high-magnetization expansion. This value has since been independently confirmed using the density matrix renormalization group method [3]. Below $q_{\min}$ the (continuous) saturation transition is governed by a remarkably simple zero-scale-factor universality [4] in which the magnetization density near saturation is described by an exactly known scaling form with no nonuniversal numbers. Using the same high-magnetization expansion it is possible to predict that there will be metamagnetism caused by the same mechanism in the unfrustrated J_1-J_2 model with an AFM first-neighbor term and an anisotropic FM second-neighbor term.

In Chap. 3 I extend this work to the two-dimensional J-Q model (see also [5]). Here I use an extension of the high-magnetization expansion from the 1D case to make a prediction for $q_{\min}$ based on an exact method. In 2D this value, $q_{\min} \approx 0.417$, converges to the infinite-size value exponentially quickly. The existence of metamagnetism is confirmed by large-scale quantum Monte Carlo simulations. For $q < q_{\min}$ the magnetization near saturation is governed by the same zero-scale-factor universality [4], but now at its upper critical dimension. We therefore expect logarithmic violations of the zero-factor scaling form at low

A. Iaizzi, *Magnetic Field Effects in Low-Dimensional Quantum Magnets*,
Springer Theses, https://doi.org/10.1007/978-3-030-01803-0_6

temperature. Detailed quantum Monte Carlo results confirm that there is indeed a low-temperature divergence, but it does not match the form predicted by [4]. I also discuss other possible forms of the divergence.

In two dimensions the zero-field J-Q model undergoes a transition from a long-range-ordered Néel phase to a valence-bond solid at $j_c \approx 0.045$. This deconfined quantum critical point is described by spinons: exotic fractionalized bosonic excitations carrying $S = \frac{1}{2}$. Previous studies [6] have suggested that a field-induçed Bose–Einstein condensate of these spinons would produce an anomalous linear temperature dependence of specific heat that cannot be explained by other mechanisms. Starting from the critical point j_c, I use an external magnetic field to induce a finite ground-state density of magnetic excitations and study these at finite temperature. I was unable to find the anomalous temperature dependence predicted by Scammell and Sushkov [6] in the spinon BEC, but at slightly higher temperatures there is a gas of spinons which does exhibit the expected anomalous temperature dependence, thus providing direct evidence for the existence of spinons. The spinon gas and the spinon BEC phases are separated by a field-induced Berezinskii–Kosterlitz–Thouless-like (BKT) transition. This BKT transition also results in a non-monotonic temperature dependence of magnetization which manifests as a finite-temperature minimum in magnetization. I discuss a rough phase boundary of this field-induced BKT transition.

References

1. A. Iaizzi, A.W. Sandvik, J. Phys. Conf. Ser. **640**, 012043 (2015). http://stacks.iop.org/1742-6596/640/i=1/a=012043
2. A. Iaizzi, K. Damle, A.W. Sandvik, Phys. Rev. B **95**, 174436 (2017). http://dx.doi.org/10.1103/PhysRevB.95.174436
3. B.-B. Mao, C. Cheng, F.-Z. Chen, H.-G. Luo, Sci. Rep. **7**, 18104 (2017). http://dx.doi.org/10.1038/s41598-017-17887-w
4. S. Sachdev, T. Senthil, R. Shankar, Phys. Rev. B **50**, 258 (1994). http://dx.doi.org/10.1103/PhysRevB.50.258
5. A. Iaizzi, K. Damle, A.W. Sandvik, Phys. Rev. B **98**, 064405 (2018). http://dx.doi.org/10.1103/PhysRevB.98.064405
6. H.D. Scammell, O.P. Sushkov, Phys. Rev. Lett. **114**, 055702 (2015). http://dx.doi.org/10.1103/PhysRevLett.114.055702

Appendix A
Supplementary Material for the 1D Few-Magnon Expansion

A.1 Few Magnons in the J-Q-h Chain

Continuing from Sect. 2.4.2, we will attempt to find $q_{\rm min}$, the value of q where the jump first appears. To do this, we will look for a direct level crossing between saturated state $m_z = S$ and the state with two flipped spins $m_z = S-2$ and therefore we must calculate $E(m_z, J, Q, L)$ for $m_z = S, S-1, S-2$. Finding energy of the saturated state is trivial: there are no places for a singlet projection operator to act, so $H\,|m_z = S\rangle = -hS$. If we add a single spin-down site (magnon), the Heisenberg term produces a tight-binding-like effective Hamiltonian on this flipped spin: the diagonal terms give it an on-site energy and the off-diagonal terms allow it to hop to neighboring sites. A Q term cannot act on this single-magnon state. The one-magnon state is a one-body problem with the analytic solution

$$E_1 = -\,J(1-\cos k) - h(S-1) \tag{A.1}$$

for periodic boundary conditions.

For purposes of algorithmic convenience, we will perform a "sublattice rotation," a unitary transformation on one sublattice which rotates $S_j^+ \rightarrow S_j^-$. This transformation has the effect of flipping the signs of all off-diagonal terms in the Hamiltonian without changing the spectrum [2]. After the sublattice rotation, Eq. (A.1) becomes:

$$E_1 = -\,J(1+\cos k) - h(S-1) \tag{A.2}$$

This appendix appears as an appendix in the article "Field-driven quantum phase transition in $S = \frac{1}{2}$ spin chains" coauthored with Anders W. Sandvik and Kedar Damle appearing in Phys. Rev. B **95**, 174436 (2017) [1]. Reprinted with permission.

A. Iaizzi, *Magnetic Field Effects in Low-Dimensional Quantum Magnets*, Springer Theses, https://doi.org/10.1007/978-3-030-01803-0

Note that the sign of the $\cos k$ term has changed. With $J > 0$, the ground state has momentum $k = 0$; therefore

$$E_1 = -2J - h(S-1) \tag{A.3}$$

for all L. For $q < q_{\min}$, the saturation field is determined by a direct level crossing between E_0 and E_1, so the saturation field is independent of Q:

$$h_s(q < q_{\min}) = 2J \tag{A.4}$$

For the two-magnon case, we can begin in the basis of the positions of each flipped spin: $|x_1, x_2\rangle$; the size of this basis is $L(L-1)/2$. We will assume that L is even. We can reduce this two-particle problem to single-particle problem using translation invariance. Consider a basis of the center-of-mass position and the distance between the spin-down sites: $|X, r\rangle$. The center of mass takes on the values $X \equiv x_2 + x_1 = 3, 4, 5, 6, \ldots (2L-1)$ and the separation takes on the values $r \equiv \min[x_2 - x_1, L + x_1 - x_2] = 1, 2, \ldots L/2$. The Hamiltonian is translation-invariant for the center-of-mass coordinate, X, so we can consider momentum states: $|K, r\rangle$, where K is the center-of-mass momentum and r is the separation between the magnons.

$$K_n = \frac{2\pi n}{L}, \quad n = 0, 1, 2, \ldots L-1 \tag{A.5}$$

For a given K_n, $r = 1, 2, 3, \ldots r_{\max}$. We must be careful with our definitions to avoid double counting states. For even-n, $r = 1, 2, \ldots L/2$, but for odd-n, $r = 1, 2, \ldots L/2 - 1$. Thus, for each of the $L/2$ even-n momentum states, there are $L/2$ r-states, and for each of the $L/2$ odd-n momentum states, there are $L/2-1$ r-states, for a total of $L(L-1)/2$ states.

Now consider how the Heisenberg term acts on a two-magnon state $|x_1, x_2\rangle$:

$$H_J |x_1, x_2\rangle = -2J |x_1, x_2\rangle - \frac{J}{2}\Big[|x_1+1, x_2\rangle + |x_1-1, x_2\rangle + |x_1, x_2+1\rangle + |x_1, x_2-1\rangle \Big]. \tag{A.6}$$

There are two ways to hop the magnons toward each other, two ways to hop them away from each other, and four ways to leave them where they are, each with magnitude $-J/2$. In the separation basis, this becomes:

$$H_J |r > 2\rangle = -2J |r\rangle - \frac{J}{2}(1 + e^{-iK}) |r-1\rangle - \frac{J}{2}(1 + e^{iK}) |r+1\rangle \tag{A.7}$$

Thus, in the "bulk" ($1 < r < L/2$), the result is very similar to the one-magnon problem. For $r = 1$, there are two slight modifications: the spin-down sites are hardcore bosons (they cannot hop across each other) and the diagonal term is only $-J$. For $r = \frac{L}{2} - 1$ and $\frac{L}{2}$, there are slight modifications due to the boundary conditions. Put this all together and we get:

$$H_J = -J \begin{pmatrix} 1 & \frac{1+e^{iK}}{2} & 0 & \cdots & & \\ \frac{1+e^{-iK}}{2} & 2 & \frac{1+e^{iK}}{2} & 0 & \cdots & \\ 0 & \frac{1+e^{-iK}}{2} & 2 & \frac{1+e^{iK}}{2} & 0 & \cdots \\ \vdots & \vdots & \ddots & \ddots & \ddots & \\ \vdots & \vdots & & \ddots & \ddots & \ddots \\ & & \frac{1+e^{-iK}}{2} & 2 & \frac{1+e^{iK}}{2} & \underline{0} \\ & & 0 & \frac{1+e^{-iK}}{2} & 2 & \underline{\frac{1+e^{iK}}{\sqrt{2}}} \\ & & & \underline{0} & \underline{\frac{1+e^{-iK}}{\sqrt{2}}} & \underline{2} \end{pmatrix} \tag{A.8}$$

where the last row and last column (underlined entries) are omitted in the odd-n momentum sectors.

Now consider the Q term, which only contributes for $r \leq 3$, so we can represent it as a 3×3 matrix:

$$H_Q = -\frac{Q}{4} \begin{pmatrix} 1 & 1+e^{iK} & e^{iK} \\ 1+e^{-iK} & 2(1+\cos K) & 1+e^{iK} \\ e^{-iK} & 1+e^{-iK} & 1 \end{pmatrix} \tag{A.9}$$

Somewhat counterintuitively, the Q term produces an effective *attractive* interaction by lowering the energy of states where the flipped spins are separated by no more than three lattice spacings. This will be the key to producing the magnetization jump.

Now, we have the energies of each magnetization sector:

$$E_S = -hS, \tag{A.10a}$$

$$E_1 = \bar{E}_1(J, Q, L) - h(S-1), \tag{A.10b}$$

$$E_2 = \bar{E}_2(J, Q, L) - h(S-2), \tag{A.10c}$$

where $\bar{E}_n$ is the ground state energy of the zero-field n-magnon chain. In order to find $q_{\min}$, we must first find the saturation field h_s by demanding that $E_S = E_2$:

$$h_s = -\frac{1}{2}\bar{E}_2(J, Q, L). \tag{A.11}$$

To guarantee a direct level crossing between $m_z = S - 2$ and $m_z = S$, require $E_1 \geq E_S = E_2$:

$$-h_s S \leq \bar{E}_1 - h_s(S-1), \tag{A.12}$$

$$h_s \geq -\bar{E}_1. \tag{A.13}$$

Combining Eqs. (A.11) and (A.13) and eliminating h_s, we find a condition for $q_{\min}$:

$$\bar{E}_2 \leq 2\bar{E}_1. \tag{A.14}$$

This condition is also essentially the condition for an attractive interaction: the energy for two magnons is *less* than twice the single-magnon energy because the interactions *lower* the total energy. From Eq. (A.2), we know that $\bar{E}_1 = -2J$, so we can find a condition on $\bar{E}_2$ for the existence of a jump:

$$\bar{E}_2 \leq -4J. \tag{A.15}$$

A.2 Derivation of the Magnetization Jump in the J_1-J_2 Chain

The anisotropic J_2 term is given by

$$H_{J_2} = -J_2 \sum_i \left[\frac{1}{4} - S_i^z S_{i+2}^z - \frac{\Delta}{2} \left(S_i^+ S_{i+2}^- + H.c. \right) \right]. \tag{A.16}$$

We will set $J_2 = -j$ ($j > 0$ is *ferro*magnetic) and follow the same steps from Sect. A.1. First, we need the one-magnon energy, which can be derived in much the same way we derived the one-magnon energy for the J-Q-h chain:

$$\bar{E}_1(j, \Delta) = -J_1(1 - \cos k) - J_2(1 - \Delta \cos 2k), \tag{A.17}$$

$$\bar{E}_1(j, \Delta) = -1 + \cos k + j - j\Delta \cos 2k. \tag{A.18}$$

Note that here we do *not* use the sublattice rotation employed in Sect. A.1; this difference can be seen by comparing Eq. (A.18), where the potential energy (-1) and kinetic energy $(\cos k)$ terms have the *opposite* sign, to Eq. (A.2), where they have the *same* sign. For $\Delta \geq 0$, $\bar{E}_1$ is always minimized by $k = \pi$. For $\Delta < 0$, $k_{\min}$ can take on two values

$$k_{\min}(j, \Delta) = \begin{cases} \pi, & (j\Delta) \geq -1/4 \\ \arccos\left(\frac{1}{4j\Delta}\right), & (j\Delta) < -1/4. \end{cases} \tag{A.19}$$

This means that the ground state energy for one magnon is given by:

$$\bar{E}_1(j, \Delta) = \begin{cases} -2 + j(1 - \Delta) & (j\Delta) \geq -1/4 \\ -1 + j(1 + \Delta) + \frac{1}{8j\Delta} & (j\Delta) < -1/4 \end{cases} \tag{A.20}$$

Now we want to write the two-magnon Hamiltonian in the separation basis (as defined in Sect. A.1). We have already worked out the separation basis Hamiltonian for the J_1 term in Eq. (A.8), but in this case we cannot use the sublattice rotation. Reversing the sublattice rotation done to Eq. (A.8), we arrive at a form for H_{J_1}:

$$H_{J_1} = J_1 \begin{pmatrix} -1 & \frac{1+e^{iK}}{2} & 0 & \cdots & & \\ \frac{1+e^{-iK}}{2} & -2 & \frac{1+e^{iK}}{2} & 0 & \cdots & \\ 0 & \frac{1+e^{-iK}}{2} & -2 & \frac{1+e^{iK}}{2} & 0 & \cdots \\ \vdots & \vdots & \ddots & \ddots & \ddots & \\ \vdots & \vdots & & \ddots & \ddots & \ddots \\ & & \frac{1+e^{-iK}}{2} & -2 & \frac{1+e^{iK}}{2} & \underline{0} \\ & & 0 & \frac{1+e^{-iK}}{2} & -2 & \underline{\frac{1+e^{iK}}{\sqrt{2}}} \\ & & & \underline{0} & \underline{\frac{1+e^{-iK}}{\sqrt{2}}} & \underline{-2} \end{pmatrix} \tag{A.21}$$

Notice that Eq. (A.21) is identical to Eq. (A.8), except for the signs of the off-diagonal terms. H_{J_2} can be derived in the same way that we derived the separation basis Hamiltonian for the Heisenberg chain, Eq. (A.8). Applying the same logic to the J_2 term, we arrive at:

$$H_{J_2}(K) = j \times \tag{A.22}$$

$$\begin{pmatrix} 2-\Delta\cos K & 0 & -\frac{\Delta(1+e^{2iK})}{2} & 0 & 0 & \cdots & \\ 0 & 1 & 0 & -\frac{\Delta(1+e^{2iK})}{2} & 0 & \cdots & \\ -\frac{\Delta(1+e^{-2iK})}{2} & 0 & 2 & 0 & -\frac{\Delta(1+e^{2iK})}{2} & \cdots & \\ & \ddots & \ddots & \ddots & \ddots & \ddots & \\ & -\frac{\Delta(1+e^{-2iK})}{2} & 0 & 2 & 0 & -\frac{\Delta(1+e^{2iK})}{2} & \underline{0} \\ & 0 & -\frac{\Delta(1+e^{-2iK})}{2} & 0 & 2 & 0 & \underline{-\frac{\Delta(1+e^{2iK})}{\sqrt{2}}} \\ & 0 & 0 & -\frac{\Delta(1+e^{-2iK})}{2} & 0 & 2-\Delta\cos K & \underline{0} \\ & \underline{0} & \underline{0} & \underline{0} & \underline{-\frac{\Delta(1+e^{-2iK})}{\sqrt{2}}} & \underline{0} & \underline{2} \end{pmatrix}$$

where the rows and columns represent $r = 1, 2, 3, \ldots L/2$. As in Sect. A.1, for even-n momentum sectors, $r = 1, 2, 3, \ldots L/2$ and for odd-n momentum sectors the basis is truncated $r = 1, 2, 3, \ldots L/2 - 1$, so we must cut off the last row and column of Eqs. (A.21) and (A.22) (the underlined entries). This approach is based on one used by Kecke et al. to study the FM-AFM J_1-J_2 chain [3].

References

1. A. Iaizzi, K. Damle, A.W. Sandvik, Phys. Rev. B **95**, 174436 (2017). http://dx.doi.org/10.1103/PhysRevB.95.174436
2. A.W. Sandvik, in *American Institute of Physics Conference Series*, ed. by A. Avella, F. Mancini, vol. 1297 (2010), pp. 135–338, http://arxiv.org/abs/1101.3281. http://dx.doi.org/10.1063/1.3518900
3. L. Kecke, T. Momoi, A. Furusaki, Phys. Rev. B **76**, 060407 (2007). http://dx.doi.org/10.1103/PhysRevB.76.060407

MIX
Papier aus verantwortungsvollen Quellen
Paper from responsible sources
FSC® C105338

If you have any concerns about our products,
you can contact us on
ProductSafety@springernature.com

In case Publisher is established outside the EU,
the EU authorized representative is:
Springer Nature Customer Service Center GmbH
Europaplatz 3, 69115 Heidelberg, Germany

Printed by Libri Plureos GmbH
in Hamburg, Germany